Managing and Marketing Radical Innovations

Marketing new technology

Birgitta Sandberg

Routledge
Taylor & Francis Group

LONDON AND NEW YORK

First published 2008
by Routledge
2 Park Square, Milton Park, Abingdon, Oxon, OX14 4RN

Simultaneously published in the USA and Canada
by Routledge
270 Madison Ave, New York NY 10016

Routledge is an imprint of the Taylor & Francis Group, an informa business

First issued in paperback 2011

© 2008 Birgitta Sandberg

Typeset in Times by Wearset Ltd, Boldon, Tyne and Wear

British Library Cataloguing in Publication Data
A catalogue record for this book is available from the British Library

Library of Congress Cataloging in Publication Data
Sandberg, Birgitta.

Managing and marketing radical innovations: marketing new
technology/Birgitta Sandberg.
p. cm. – (Routledge studies in innovation, organization and technology; 8)
Includes bibliographical references and index.

1. Technological innovations–Marketing. 2. Technological
innovations–Management. 3. New products. I. Title.

HD45.S272 2008
658.4'063-dc22

2007035793

ISBN10: 0-415-43307-X (hbk)
ISBN10: 0-415-61947-5 (pbk)
ISBN10: 0-203-93048-7 (ebk)

ISBN13: 978-0-415-43307-5 (hbk)
ISBN13: 978-0-415-61947-9 (pbk)
ISBN13: 978-0-203-93048-9 (ebk)

Managing and Marketing Radical Innovations

This book responds to a growing demand in the academic community for a focus on customer-related proactive behavior in the study of radical innovation development, combining a thorough theoretical discussion with detailed international case studies considering the role of this proactivity in five firms engaged in the process.

Unlike other studies in this area, this book demonstrates that anticipation plays an important role at the idea-generation stage, and Sandberg introduces a new way of describing a firm's proactivity as a dynamic pattern. Furthermore, the deeper consideration of customer-related proactivity contributes to the study of market orientation, which increasingly focuses on the proactive side.

This book will be of great interest to students and researchers engaged with radical innovation and market creation as well as those taking courses in Innovation Management.

Birgitta Sandberg is Assistant Professor in International Business and coordinator of the Global Innovation Management Masters Degree Programme at the Turku School of Economics, Finland.

RIOT!

Routledge studies in innovation, organization and technology

Contents

Figures

Tables

Preface

This book combines two dynamic and multifaceted themes: the development of radical innovations and proactiveness, both subjects of growing interest to researchers and managers. In order to cope with the fast-changing environment, tightening competition and customers who are more demanding, many firms strive to produce increasingly radical innovations. Even though radical innovations may lead to – at least temporary – competitive advantage, their creation is extremely risky, and the failure rates are high.

One reason for the failure is that firms developing these innovations may be facing "a mission impossible." On the one hand, in order to succeed, they have to develop innovations that fulfill the needs of customers. This is well in accordance with the old marketing mottos: "Know your customers!" "Listen to your customers!" "Satisfy the needs of your customers!" With radical innovations, however, it is not that easy: the market does not yet exist. Firms may not even know who the prospective customers might be. Furthermore, even if the customers are known, they often do not recognize the need for a completely new kind of product or service. It is thus impossible for the firms to react to customer needs. The market may be hidden – even from the firm that is creating it. For this reason, proactive behavior toward customers, in other words anticipating and influencing them, may be the key issue in building a bridge between the innovation and the market.

Managing and Marketing Radical Innovations reports on a recent research project involving a thorough theoretical discussion on customer-related proactiveness and radical innovation development and detailed case descriptions concerning the role of this proactiveness in five Finnish firms engaged in the process. Each of the case studies tells a fascinating success story. How was the firm able to create a market for its radical innovation?

Finnish companies are particularly interesting to study because Finland has performed extremely well in global competitiveness rankings, and it also ranks among the top Organization for Economic Cooperation and Development (OECD) countries in terms of technology development. Furthermore, due to the small size of the home market, Finnish firms have been forced to give considerable attention to their international customers in their innovation strategies. Thus, one could argue that they have learnt to cope with globalization pressures earlier than firms originating from larger countries.

The study contributes to previous literature in that it is among the first to link customer-related proactiveness with radical innovation development. It illustrates how the degree of proactiveness toward customers changes during the process. Unlike many earlier studies, it demonstrates that anticipation seems to play an important role at the idea-generation stage. The research also elaborates the concept of proactiveness and introduces a new way of describing a firm's proactiveness as a dynamic pattern. Furthermore, the deeper consideration of customer-related proactiveness contributes to the study of market orientation, which increasingly focuses on the proactive side. This book is therefore a response to recent requests for academics to pay more attention to proactive behavior, radical innovation development and the international scope of innovation creation.

Acknowledgments

This book is the outcome of a long research project. I am almost ashamed to admit that most of the time I have had fun. For the pleasure I have had on this journey, I am indebted to a large number of people and institutions to whom I would like to express my gratitude.

First of all, I would like to thank Professor Emeritus Sten-Olof Hansén and Professor Niina Nummela for their encouragement and insightful comments along the whole research process. Furthermore, various special people at the department of marketing have given me valuable advice. My sincere thanks are due to Professors Karin Holstius, Urpo Kivikari, Leila Hurmerinta-Peltomäki, Kari Liuhto and Jorma Taina for their support. I am also grateful to Dr. Zsuzsanna Vincze, with whom I have had many enjoyable discussions about radical and not-so-radical issues in life.

Alongside those at the Turku School of Economics, people from other universities have also left their fingerprints on my work. I have had the privilege of receiving comments from Professor Hans Georg Gemünden (Technische Universität Berlin), Docent Arto Rajala (Helsinki School of Economics), Professor Kjell Grønhaug (The Norwegian School of Economics and Business Administration), Professor Kristian Möller (Helsinki School of Economics) and Professor Anthony Di Benedetto (Temple University). Thank you all for your kind advice!

In the very early stages of the process, Professor Bill McKelvey (UCLA Anderson School of Management) and Professor Stanley W. Mandel (Wake Forest University) gave me much encouragement. Later on, Professor Olli-Pekka Hilmola (Lappeenranta University of Technology) provided me with important references and helped me with the online survey. Thank you, one and all!

I am indebted to Sari Digert, Elina Halonen, Auli Rahkala and Eeva Kettunen, who have assisted me with the preparation of this book. I also wish to extend my warmest thanks to Joan Nordlund, M.A., for her devotion to making my English fluent.

I am grateful to the Publication Board of the Turku School of Economics for giving me permission to modify this book from the dissertation published earlier under the title *THE HIDDEN MARKET – EVEN FOR THOSE WHO CREATE*

IT? Customer-Related Proactiveness in Developing Radical Innovations. I would also like to thank the publishers for generously granting me permission to use excerpts from the following publications:

- Birgitta Sandberg, 2004, Creating an International Market for Disruptive Innovation. The Domosedan Case. In: *In Search of Excellence in International Business. Essays in Honour of Professor Sten-Olof Hansén.* Ed. by Niina Nummela, 71–95. Publications of the Turku School of Economics and Business Administration. Series C 1:2004. Turku. Used by permission of Turku School of Economics.
- Birgitta Sandberg and Leena Aarikka-Stenroos, 2005, *Did you forget your Skis at Home? How to Turn "Silly-Sticks" into Internationally Fashionable Sports Equipment?* Publications of the Turku School of Economics and Business Administration. Series B-2:2005. Turku. Used by permission of Leena Aarikka-Stenroos and Turku School of Economics.
- Birgitta Sandberg, Creating the Market for Disruptive Innovation: Market Proactiveness at the Launch Stage, 2002, *Journal of Targeting, Measurement & Analysis,* Vol. 11, No. 2, 184–196. Henry Stewart Publications, reproduced with permission of Palgrave Macmillan.
- Birgitta Sandberg, Customer-Related Proactiveness in the Radical Innovation Development Process, 2007, *European Journal of Innovation Management,* Vol. 10, No. 2, 252–267. Used by permission of Emerald Group Publishing Limited.

For granting permission to use the illustrations in the case descriptions, I would like to thank Finnzymes Ltd, Itella Corporation, Exel Sports Ltd, Marioff Corporation and Suunto Ltd. I also wish to express my appreciation and thanks to the various interviewees who gave me their precious time and openly shared their experiences of innovation development. I am particularly indebted to Johannes Koroma, who made the survey possible. The Foundation for Economic Education, the Paulo Foundation and the Turku School of Economics provided the financial support that made the research possible. I am deeply grateful.

Finally, I wish to extend my warmest thanks to the people closest to me, my husband Jarkko and my mother Seija, for their patience, encouragement and support.

Birgitta Sandberg

1 Introduction

1.1 Creating the market for radical innovations

1.1.1 The hidden market for radical innovations

Radical innovations are innovations that are new both for the firm and the market. According to Stefik and Stefik (2006: 3), "they do something that most people did not realize was possible." It has been claimed that their development is critical to the long-term survival of many firms, since they provide the foundation on which future generations of products or services are created (McDermott and O'Connor 2002: 424). Furthermore, previous studies (e.g. Chaney *et al.* 1991) indicate that radical innovations have a greater return on investment than less-original new products. Not only do they have an impact on firms, they also affect society and customers, since they are engines of economic growth and sources of better products (Chandy and Tellis 2000: 2).

Although the importance of radical innovations is widely recognized, developing them is still rather poorly understood (Leifer *et al.* 2001: 102). Further research is thus needed, especially since the failure rate is particularly high due to the various challenges inherent in their development (cf. *New Products Management for the 1980s* 1982: 9).

The newness of radical innovations creates various challenges for the innovating firms. First, when transforming an idea into an invention, firms need to be able to cope with significant technological uncertainty. Overcoming the technological challenges is not enough to turn invention into innovation, i.e. to make it succeed commercially, however. Commercializing a radical invention also requires coping with considerable market uncertainty (cf. Dougherty 1996: 426–427; McDermott and O'Connor 2002: 430; Veryzer 1998a: 306–307).

Past literature emphasizes the fact that inventions can be turned into innovations when they fulfill customer needs (e.g. Crawford 1977; Holt *et al.* 1984: 27). It has been suggested that firms developing innovations *ought to pay attention to the customers and their needs throughout the development process*, from idea generation to launch (Holt *et al.* 1984: 1). This includes seeing marketing as dispersed in various functions within the firm and not as a separate function (cf. Moorman and Rust 1999: 180).

Creating a radical innovation that fulfills customer needs is extremely difficult, however, since at the earliest stages of the development process the firm *may not even know who the customers*[1] *for the innovation under development might be* (Deszca *et al.* 1999: 617). Furthermore, even if the customers are known, they themselves often *cannot articulate the need* for a completely new kind of product or service (Hoeffler 2003: 411; Veryzer 1998b: 143). Accordingly, in the case of radical innovations, "user-needs emerge only gradually for the users themselves" (Hyysalo 2004: 208; cf. Berthon *et al.* 2004: 1067). This may even make the collection of customer feedback insignificant and the results distorted. It has been argued that focusing on customers may even impede radical innovations (Christensen 1997: 208).

In sum, due to the inherent newness, it may be extremely difficult to react to customers' needs and wishes during the radical innovation development process, even though reactiveness would often increase the chances of success in the marketplace. How, then, are markets for radical innovations created? Is some kind of reaction still possible? Are the firms developing these kinds of innovations able to anticipate the needs or even influence them? In fact, proactive behavior toward customers in terms of anticipating and influencing their needs may play an important role in building bridges between the innovation and the market.

1.1.2 *The role of proactiveness in market creation*

Proactive behavior, i.e. anticipating changes and even altering the environment, is often seen as a way in which firms can survive and actually benefit from the changes in their environment[2] (Harper 2000: 75–77; Morgan 1992: 24–37). However, proactiveness does not necessarily lead to success. It has been pointed out in studies by Tan and Litschert (1994: 13) and Liuhto (1999: 105), for example, that there are also situations in which reactive behavior brings better results.

Some earlier studies on innovation management have highlighted the importance of proactive approaches when developing innovations. Kaplan (1999) and Rice *et al.* (1998) accentuated the role of proactiveness in stimulating radical innovations within the firm. Hyysalo's (2004) study emphasized the importance of anticipating prospective use during a radical innovation development process, and the study by O'Connor and Veryzer (2001) concentrated on market visioning in radical innovation development.

Narver *et al.* (2004) studied reactive and proactive market orientation in terms of creating and sustaining new-product success and came to the conclusion that both were needed, but proactive orientation seems to be especially important. Furthermore, Verganti (1999) studied how anticipation and reaction interact in the development of radical innovations. He concentrated only on the early phase of product development – the phase before the start of the detailed product design. He identified four different approaches to combining anticipation and reaction and suggested that none of these was a best practice in itself.

He also found that anticipation and reaction were not mutually exclusive but rather interacted with each other. Early planning for reaction measures was considered important. These results are in accordance with those reported in a study by Evans (1991: 78–83) indicating that both proactive and reactive maneuvers are important for high-technology firms.

All in all, there are a very limited number of studies that concentrate on proactiveness and even fewer that concentrate on customer-related proactiveness in developing radical innovations. Further, there seems to be a lack of studies dealing especially with proactiveness during the whole innovation development process or concentrating on how to influence or create proactiveness in this process. This is rather surprising, since anticipating changes and influencing the environment seem to be crucial in developing radical innovations because products, markets and competitive boundaries are in a state of flux at that time (cf. Evans 1991: 68).

Firms creating radical innovations are frequently cited as examples of proactiveness (e.g. O'Connor 1998: 152; Salavou and Lioukas 2003: 103). However, the overall proactiveness of these firms may be questioned, since behavior toward customers may still be quite conventional, limited to selling the ready product to those who might not even know what they want (cf. Workman 1993a). Hence, these firms may behave proactively in the technical aspects of product development, but still be reactive toward their customers.

Nevertheless, in situations in which customer needs cannot be reacted to, proactive behavior toward them may enable firms to link new inventions to market opportunities (e.g. Nayak and Ketteringham 1986). Thus, it seems especially important to study customer-related proactiveness during the development of radical innovations more carefully. Since radical innovations are often targeted at international markets (e.g. Acs *et al.* 2001), the international aspect of the development cannot be ignored.

1.2 The globalization of radical innovation development

1.2.1 The utilization of knowledge on the global scale

Research on how globalization is affecting radical innovations is scarce. However, on the basis of findings from the few studies carried out on its effect on innovations in general (e.g. Acs *et al.* 2001; Devinney 1995; Gerybadze and Reger 1999; Golder 2000b), it could be concluded that it is redefining the nature of radical innovations as well. According to Devinney (1995: 74), since globalization is increasing the amount and diversity of information, and because the management of the innovation process is almost synonymous with the management of information acquisition, it is obvious that globalization is likely to lead to more innovation opportunities. However, realizing these opportunities also demands the effective dissemination and utilization of that information (cf. Kohli and Jaworski 1990: 3).

According to the study carried out by Gerybadze and Reger (1999: 264–266), distributed research and development activities and globally dispersed innovation processes are driven by the need to secure access to the most critical resources.[3] One of the most critical resources for radical innovations is knowledge. The knowledge needed in innovation development could be categorized as R&D-based and market-based. *R&D-based knowledge* regarding the creation of the radical innovation is new raw information and thus is often communicated through face-to-face communication[4] (see Meyer 1995: 180–181). Consequently, especially in R&D-intensive industries, the innovative activity tends to cluster spatially (Audretsch and Feldman 1996). This creates pressures toward concentration of the whole radical innovation development process in one location.

On the other hand, global competitiveness often requires tapping sources of technology around the world (Meyer 1995: 194). The extent of the internationalization of R&D activities has increased considerably since the 1980s (Gerybadze and Reger 1999: 254). Large multinational firms have been the drivers of this trend by extending their R&D to different foreign locations, which has led to the creation of multiple centers of knowledge in several geographical locations. Consequently, firms that strive to be among the leading innovators need to be present in more than one of these prime centers of science and technology. This global sourcing in the area of research and technology means that corporate learning can take place in various distributed knowledge centers (Gerybadze and Reger 1999: 254–255). Sölvell and Zander (1995: 21–23) use the concept "selective tapping" to describe how the firm creates its core technology in the home base and simultaneously use subsidiaries to tap the international environment to support this creation. They emphasize the fact that, even though it is rather easy to source basic factors of production worldwide, it is much more difficult to tap advanced factors and tacit knowledge from abroad.

The emphasis on access to R&D-based knowledge has lately been broadened to cover access to *knowledge about lead markets*. Presence in lead markets and learning from them is becoming increasingly important for innovative firms, as comparative advantage is apparently built on the efficient coupling of lead marketing, R&D and innovation (Beise 2004; Gerybadze and Reger 1999: 264–271). Nevertheless, this coupling on the international scale is rather demanding. Golder (2000b) studied the innovation development process in international markets and came to the conclusion that, although firms recognize the importance of sharing market intelligence, they admit that it is not done sufficiently on the international scale.

In any case, although a presence in the knowledge centers and/or lead markets can enhance the firm's absorptive capacities[5] (Cohen and Levinthal 1990; Holden 2002: 251–253), these capacities are also increasingly acquired externally through cooperation and networks (cf. Ritter and Gemünden 2003). The extent of this external procurement depends on both internal resources and transaction costs (cf. Nooteboom 2000).

However, since the studies presented above did not particularly focus on radical innovations or customer-related proactiveness, these findings are of limited use for this study. Although the dependence of the innovation development process on external knowledge exchange has been widely acknowledged (e.g. Golder 2000b; von Hippel 1986; Zander and Sölvell 2000), the results of studies on radical innovations indicate that their newness may considerably hinder the utilization of external knowledge (Christensen 1997; Veryzer 1998b). Nevertheless, it would be especially important to be able to apply external knowledge in the development of radical innovations, which often tend to be targeted toward international markets.

1.2.2 Targeting global markets

The international perspective also seems to be important in the search for markets for radical innovations. According to Acs *et al.* (2001), firms creating them almost always have to go international. It is extremely difficult to operate locally and simultaneously to find protection from intellectual piracy from abroad, and it is thus of paramount importance to be able to introduce innovations in many countries simultaneously. (Acs *et al.* 2001: 239; cf. Chryssochoidis and Wong 1996: 179; Millier 1999: 32).

In any case, large markets are needed to cover the high cost and risk associated with the development of a radical innovation (cf. Devinney 1995: 72). The need for international markets is particularly prevalent in situations in which the firm operates in narrow niches that are scattered thinly around the world (Hurmelinna *et al.* 2002).

According to Oesterle (1997), the internationalization of firms producing radical innovations is dependent on the climate in the home market. If the climate is hostile, i.e. buying power is low, the acceptance rate is generally low and competitors are defending their market share aggressively, the radical innovation is immediately taken to international markets. However, if the climate is friendly, radical innovations are first launched in the home market, and the internationalization follows the sequential process of market entry (Oesterle 1997: 141–144).

Even though radical inventions are often created in small firms, multinationals tend to have advantages in terms of their global commercialization because they are able to quickly introduce them in various countries simultaneously. Small firms, on the other hand, may lose control over their intellectual property, since they are often forced to use licensing when introducing their innovation internationally, for example. Since the risk of intellectual piracy is particularly accentuated in the case of radical innovations, it has been suggested that small and large firms could act together as symbiotic partners in creating global markets for them (Acs *et al.* 2001: 239–241).

In sum, the globalization of markets seems to put more pressure on global-scale market proactiveness: competitors can turn up anywhere at any time. It is also causing worldwide convergence of market structures (Levitt 1983) and thus creating more possibilities for international market proactiveness. The significance of international markets is generally recognized in innovation management, although most studies deal with incremental innovations and the coordination of their development in multinational firms (e.g. Golder 2000b; Ridderstråle 1996). Radical innovations are not usually considered, nor is the variety in size of the firms creating them. *Hence, there seems to be a lack of studies on the presence of the international context in the radical innovation development process.*

1.3 The research setting

1.3.1 The purpose of the study

Various researchers (e.g. John *et al.* 1999: 78; Workman 1993a: 405) have stated that there is a need for studies that concentrate on the linkage between innovation and the customers. Drucker (1974: 64–67) even argued that the two essential activities of business are "innovation" and "marketing," and both of them are needed for long-term success.

However, as stated in Section 1.1, coupling innovation development with customer needs seems to be particularly difficult in the development of radical innovations. Their newness hinders reaction to customer needs. Therefore, customer-related proactiveness (i.e. acting based on the information gathered about the customers before their behavior has had a direct impact on the firm or deliberately influencing and creating changes in customer behavior) may play an important role in creating a market that did not even exist when the innovation development began.

The purpose of this study is to analyze the role of customer-related proactiveness in the process of developing radical innovations. In order to achieve this understanding, it is first necessary to deliberate on the concept of customer-related proactiveness more carefully, thereby enabling analysis of its degree and international scope along the innovation development process. Furthermore, since it has been argued that understanding an organizational process includes understanding the context (cf. Pettigrew 1988), the role of organizational variables in the creation of proactiveness ought to be considered as well. Consequently, the purpose of the study could be further divided into the following sub-objectives:

1   to analyze what customer-related proactiveness in developing radical innovations encompasses;
2   to analyze the degree of customer-related proactiveness along the radical innovation development process;
3   to analyze the manifestation of the international context in customer-related proactiveness during the radical innovation development process;
4   to describe how customer-related proactiveness is created in firms producing radical innovations.

This study is limited to firms aiming at mastering innovation creation as a whole: they create the ideas, do the development work and finally launch the innovation. Thus, the emerging reconnaissance industry, i.e. the array of firms that constantly imagine opportunities and generate new ideas but let other firms take the inventions to the market (e.g. through licensing), is not included.

The theoretical contribution of the focal study lies in providing an analysis of customer-related proactiveness in general, and especially in its role in developing radical innovations. Furthermore, it is also likely to provide better analytical tools for such an endeavor. Although the process of innovation development has been widely studied, and lately there has been a growing stream of research on radical innovations in particular, this study seems to be among the first to suggest a link with customer-related proactiveness.

One criterion for selecting the research topic was that the study should provide some *managerial implications* for innovative firms. Highly innovative firms may find that customer-related proactiveness is the crucial tool for shaping science into a commercial product. It is therefore assumed that innovative firms, which are facing increasing pressure to enhance their proactiveness toward customers, need more knowledge on how it can be fostered in the organization.

1.3.2 The research strategy

The purpose of the study, i.e. to analyze the role of customer-related proactiveness in the process of developing radical innovations, demands a deeper *understanding* of the phenomenon. This naturally guides the methodological choices for the empirical research. In this study, the understanding is gained through *describing and analyzing* the degree and international scope of customer-related proactiveness (content) in the different phases of radical innovation development (the process), taking into account the contributory roles of the organization and the environment (the context) (cf. Pettigrew 1992). Since there is no clear theoretical guidance for studying customer-related proactiveness, and the theoretical background has to be built up from diverse streams, the study has *exploratory* elements.

The research is realized through a process of gradual framework building. The initial framework describes how customer-related proactiveness is created and how it changes along the radical innovation development process. Figure 1.1 illustrates how the initial framework is based on theory, justified by the fact that there are plenty of existing studies on the development of radical innovations, the marketing of innovations and proactive behavior. Thus, it seems rational to use existing work as a basis for this study (cf. Miles and Huberman 1994: 17). Conceptual discussion plays an important role in the framework, since the key concept of the study, "customer-related proactiveness," has not previously been clearly defined, and even the concept of "proactiveness," although frequently used, is rarely defined in the literature.[6]

Although the presence of a preliminary theoretical framework inevitably means that some issues are disregarded, while others arise, its utilization could be justified for a number of reasons. First, although there are few studies specifically dealing with customer-related proactiveness during the radical innovation

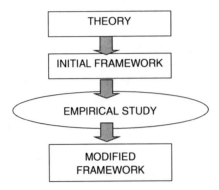

Figure 1.1 Framework building in the study.

development process, there are plenty of separate ones on the development of radical innovations, the marketing of innovations and proactive behavior. Thus, it seems rational to incorporate existing work into the basis of this study (cf. Miles and Huberman 1994: 17; Ghauri and Grønhaug 2002: 49). Furthermore, studying a processual phenomenon (innovation development in this case) is very difficult without a theoretical framework that helps the researcher to focus the analysis and to see the connections between context and the content (Odén 1989: 138–139; Halinen 1994: 32; Pettigrew 1990: 272). In any case, an explicit framework links the study to the existing research and in that way serves the theory building (Normann 1973: 54–55). It has also been suggested that a priori theoretical know-ledge of the subject area increases the likelihood of being able to generalize the results of case studies to larger contexts (Lukka and Kasanen 1995: 84).

Miles (1979: 591) argued that, even when only a rough preliminary frame-work can be formed before the fieldwork begins, it is still often beneficial as long as it can be revised along the way. In this case, the a priori framework is used to guide and focus the investigation and to facilitate comparison between the radical innovations under study. Since it is a preliminary attempt to combine the findings of past studies in order to describe proactiveness during the radical innovation development process, it does not provide a good-enough basis for hypothesis testing. On the contrary, *the framework is left open for modification based on what emerges from the empirical data.* Carson *et al.* (2001: 11–12), among others, support this kind of procedure, arguing in favor of balancing the inductive and deductive approaches. They suggest a framework derived from a literature analysis, which is then evaluated empirically by allowing new insights to emerge.

The research strategy adopted is that of the *field study*, since the focus is on realistic processes of radical innovation development, i.e. on particular behavior systems analyzed by means of unobtrusive research operations (cf. McGrath 1982: 72–75). The field study is carried out according to case-study methodol-ogy, and the initial framework is linked to the empirical cases by applying pattern-matching logic. Finally, the framework is modified based on the results of the case study.

A case-study strategy is chosen to further understanding of a firm's proactiveness toward its customers because it allows holistic understanding of complex phenomena that cannot be easily separated from their organizational context (cf. Pihlanto 1994: 373; Yin 1981: 59). According to Valdelin (1974: 46–47), case-study research enables us to study various aspects, examine them in relation to each other and contemplate the phenomenon within its environment. Eisenhardt (1989: 534) further states that it "focuses on understanding the dynamics present within single settings." It characteristically takes a comprehensive conceptual framework and confronts it with a complex set of data at a large number of points simultaneously (Normann 1973: 50). It enables the usage of data triangulation, which helps the researcher to achieve a more holistic understanding of the phenomenon (Ghauri 2004: 115–116; Yin 1989: 95–97). Thus, the case-study method maximizes the realism of the context at the expense of precision and generalizability (cf. McGrath 1982: 72–75).

This strategy also seems to facilitate study of the whole radical innovation development process: the importance of studying a process in its context,[7] i.e. embeddedness, has been emphasized by Pettigrew (1990: 269; 1992: 10), for example. A further justification is that the number of radical innovation development processes to be studied is rather limited. The unit of analysis, *"a case" in this study, is seen as a development process of a particular radical innovation.* The cases are instrumental (cf. Stake 1995), i.e. they serve as a tool for enhancing our understanding about customer-related proactiveness during the development of radical innovations.

Since the study focuses on the innovation development *process*, it is inherently *longitudinal*. A process could be regarded as "a change, transition, or transformation of some entity over time" (Halinen and Törnroos 1995: 501), hence the study of sequences of events is a prerequisite of process analysis. However, it is not only a question of describing the sequences, it also involves the identification of patterns (Pettigrew 1992: 8) and the search for underlying mechanisms that shape the patterning (Pettigrew 1997: 339). The search is not for linear explanations but rather for holistic explanations, i.e. intersecting sets of conditions that link features of the process and its context to certain outcomes (Pettigrew 1997: 341–342; Ragin 1987: 23–30).

Longitudinal study is based on either real-time or retrospective data. This study relies solely on *retrospective* data and could thus also be characterized as *historical* research. Although it would be better to observe the innovation development processes throughout its unfolding (cf. Millier 1999: 23–24; Van de Ven and Poole 1990: 316–317), this is not a very realistic option as the study focuses on radical innovations. In other words, within the scope of one small research project, it would be practically impossible to follow the development processes of hundreds of radical ideas over many years[8] to find out if one (or any) of them turns into a commercially successful product or service (i.e. innovation). After all, success can only be assessed ex post.

1.4 Positioning the study

Several theories were trawled in order to find a suitable theoretical background for the study. The criteria for selecting a theory/theories that matched its purpose were set as follows:

1   The theory should provide a solid base for analyzing the content of proactive behavior. This implies a relatively long research tradition as well as the potential application to the analysis of organizational behavior in turbulent circumstances.
2   The theory has to explain customer-related proactiveness on the firm level and not focus on individual decision-making or industry-level strategic behavior.
3   The theory should also provide the grounds for analyzing the underlying dynamics and variables that influence customer-related proactiveness.

These criteria seemed to turn attention toward the vast and multifaceted constellation of organizational theories. Marketing theories, although having a long and rich research tradition in terms of the effects of environmental factors on marketing and vice versa,[9] seemed to be too limited as such, given the scope of the current study, i.e. proaction. Thus, attention was first focused on organizational theories and then on marketing theories.

1.4.1 Organizational theories

Astley and Van de Ven (1983) classified organizational theories according to micro- and macro-levels of organizational analysis and deterministic versus voluntaristic assumptions of human nature. The result is the emergence of four basic perspectives: natural selection, collective action, system structural and strategic choice (see Figure 1.2).

Since this study aims at analyzing behavior at the firm level, the interest is naturally in the theories presented in the figure on the micro-level. The next task

	DETERMINISTIC ORIENTATION	VOLUNTARISTIC ORIENTATION
MACRO-LEVEL	Natural selection view	Collective-action view
MICRO-LEVEL	System-structural view	Strategic choice view

Figure 1.2 A classification of organizational thought (adapted from Astley and Van de Ven 1983: 247).

is to choose between a deterministic and a voluntaristic orientation. The former presupposes that exogenous forces determine human beings and their institutions, and the focus is on the structural properties of the context within which the action unfolds. Individual behavior is seen as determined by the structural constraints that provide organizational life. The voluntaristic orientation puts individuals as the basic unit of analysis and source of change in organizational life. (Astley and Van de Ven 1983: 246–247; Bourgeois 1984: 590) Since one of the criteria is to focus on the firm level rather than individual decision-making, the deterministic orientation seems to be more appropriate, *provided that* it is able to offer a broad enough theoretical base for analyzing proactive behavior. At first sight this does not seem possible: Astley and Van de Ven (1983) pointed out that the manager's role in this case is considered reactive rather than proactive as in cases of strategic choice. However, it has also been argued that the concept of proactivity also includes the seeds of environmental determinism, since the firm is always proactive toward something (cf. Vesalainen 1995: 31). In other words, proactiveness, just like reactiveness, inherently requires an objective: proactive toward what? Thus it appears that, after all, organizational thought with a deterministic orientation does not exclude the analysis of proactiveness.

It thus follows that the *system-structural view* is the most appropriate for this study. However, this does not mean neglecting the role of the decision-makers. Even if the emphasis is on the firm level, the role of managers obviously also has to be taken into account since firms respond mostly to what their managers perceive (Miles and Snow 1978: 20).

According to Astley and Van de Ven (1983: 247), the system-structural view comprises the following schools of organizational thought: structural functionalism, contingency theory and systems theory. Structural functionalism emphasizes the role of structural elements in the achievement of organizational goals. The behavior of organizations is seen to be very rational and constrained by externalities, and human actions are seen to be determined by the institutional structures Thus, the emphasis is on reactiveness (cf. Parsons 1951). Structural functionalism seems to be too narrow a concept to form the theoretical basis of the study: proactiveness is not taken into account and the strategic choice of managers is completely neglected (see e.g. Donaldson 1997).

Contingency theory and systems theory (especially *open systems theory*) were thus selected as the basis of the study, since they fulfill all the criteria mentioned at the beginning of this section. They take account of the open character of systems, facilitate the analysis of both reactive and proactive behavior and consider the dynamic changing nature of systems. The use of both theories is justified, since they apparently complement each other, and both have their roots in self-organizing-systems theories. Moreover, both have a long research tradition behind them: contingency theory has long been used in addressing concepts such as "reactive" and "adaptive," while open-systems theory seems to offer more grounds for analyzing proactive behavior and changes in proactiveness.

However, in order to understand proactiveness toward *customers*, contingency and open-systems theories need to be combined with marketing thought.

The usability of different streams of marketing research is evaluated in the following section.

1.4.2 Marketing theories

Although the main theoretical background for analyzing proactiveness is in organizational theories, the marketing dimension ought not to be forgotten in the theoretical discussion. Many writers (e.g. Savitt 1987: 308; Zeithaml and Zeithaml 1984: 46) have criticized the reactive orientation that marketing thought seems to have adopted toward the environment. The notion of firms as creators and shapers of their environment emerged in the literature on marketing strategy in the 1980s (Zinkhan and Pereira 1994: 192–194). This body of studies seems highly relevant here, particularly in terms of the discussion on the concept of customer-related proactiveness. In this context, studies on *market orientation* have played a major role (cf. Homburg and Pflesser 2000; Zinkhan and Pereira 1994: 194).

Market-orientation research[10] can be divided in two streams: behavioral and cultural. From the behavioral perspective, it is considered in terms of specific behaviors (Homburg and Pflesser 2000: 449; Kohli and Jaworski 1990), whereas the cultural perspective refers to certain types of organizational culture (Homburg and Pflesser 2000: 449; Narver and Slater 1990: 21). This study is more closely related to the behavioral stream of research, since the focus is on proactive behaviors. Nevertheless, these behaviors stem from the organizational culture (cf. Homburg and Pflesser 2000: 451) that belongs to the organizational variables influencing the creation of proactiveness.

Studies that combine marketing with the development of radical innovations are also of relevance here: they fall into three classes. The first class comprises *studies that concentrate on the market–technology link* and apparently acknowledge the special features inherent in the marketing of radical innovations. John *et al.* (1999) identified special features of the technology-intensive market and their impact on marketing decisions. Further, Gardner *et al.* (2000) studied the special nature of marketing high-technology products and came to the conclusion that the marketing strategy differs substantially from the marketing of low-technology products. A study conducted by Roberts (1990) found that the market–technology linkage evolves: the market orientation of the technology-based firms in question increased over time. Millier (1999: 3) studied the industrial marketing of radical technological innovations and concluded that the limited role of marketing in innovation development was due to the fact that it had traditionally been designed for situations in which there were products to manage and markets to analyze. However, as he also argues, it is not feasible to wait for the innovation to go to the market before starting to market it (Millier 1999: 34).

Second, there is a vast array of *studies that concentrate on the specific marketing tactics used during certain phases of innovation development*. In general, the findings seem to be rather contradictory. Hayes and Abernathy (1980: 71) proposed that the adoption of a market-driven strategy, with its emphasis on customers as the sources of new product ideas, is unlikely to lead to breakthrough innovations. This statement was further supported in the empirical research of Workman (1993a and 1993b), Atuahene-Gima (1996) and O'Connor and Veryzer (2001). On the other hand, von Hippel *et al.* (2000: 22) suggested that identifying lead users and learning from them should be taken as a systematic approach to generating radical innovations.[11] Lettl *et al.* (2006) obtained similar results when they studied radical innovation development processes in the field of medical technology. Thus, the role of customers in the development of radical innovations is open to question.

Many studies concentrate on the launch phase of the innovation (e.g. Easingwood and Koustelos 2000; Mullins and Sutherland 1998; Rackham 1998; Urban *et al.* 1996). Lynn *et al.* (1996) studied how firms developed an understanding of the markets for radical innovations and found out that probing and learning played an important part in this process. Consequently, instead of analyzing the market in detail and carefully selecting the best alternative, these firms seemed to introduce an early version of the innovation, to learn from the feedback and to make modifications accordingly (Lynn *et al.* 1996). A later study by Mullins and Sutherland (1998) supported these findings. However, according to O'Connor (1998), the role of market learning is not so apparent in the new product development projects concerning radical innovations, and probing and learning is only one of the mechanisms that are used to manage market uncertainty.

Examples of more specific research areas include Goffin's (1998) study of the ways in which high-tech firms evaluate customer support during new-product development and Haverila's (1995) dissertation on the role of marketing when new high-technology products are launched onto international markets. His study concentrated on the usage of different marketing methods (e.g. personal selling, advertising, differentiation, marketing consultants) and their effectiveness.

Third, there is an increasing stream of *studies on cooperation in the marketing of radical innovations*. The topics they cover include intra-organizational relationships between the different marketing units (Möller and Rajala 1999), cross-functional cooperation (Gupta *et al.* 1986; Jassawalla and Sashittal 1998; Rajala 1997; Shanklin and Ryans 1984; Song *et al.* 1997), and inter-firm cooperation (Atuahene-Gima and Evangelista 2000; Teece 1992). The increase in the number of these studies is related to the fact that marketing activities are becoming more and more demanding in high-technology firms due to the special characteristics of products (intensified technological complexity, high knowledge intensity and systemic features). This increases the level of tacit knowledge involved in both product development and marketing. Hence, marketing activities are more often spread out among various organizational units, and the communication between them has become increasingly important (Möller and Rajala 1999; Olson *et al.* 1995).

1.4.3 Synthesis

In sum, a considerable variety of studies connect marketing to the development of radical innovations. Naturally, the three different streams mentioned above are closely interconnected. The market–technology linkage is reflected both in the specific tactics used in the different phases of innovation development and in the marketing cooperation. Since this linkage seems to be the underlying force, it is also of greatest interest here, although it would be rather difficult to study it without considering what it reflects in both tactics and cooperation. Since proactiveness could also be contemplated on the tactical level (see Section 3.1), tactics seem to be a more natural choice for this study.

Consequently, the study does not explicitly focus on the role of cooperation in the marketing of radical innovations. Cooperation is regarded only in as far as it is necessary in order to understand how it may contribute to the firm's customer-related proactiveness. This limitation seems to be necessary in order to direct attention toward the focal phenomenon, i.e. customer-related proactiveness.

In sum, Figure 1.3 illustrates the theoretical positioning of the study. Although *customer-related proactiveness* in itself has not been extensively investigated, there is a vast amount of literature on proactive behavior, on the

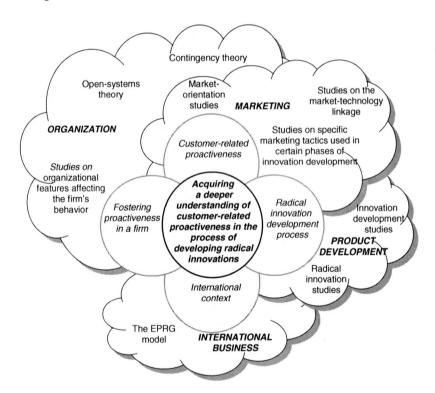

Figure 1.3 The purpose, key concepts and positioning of the study.

one hand, and on market orientation, on the other. These two streams are combined in the theoretical framework of this study with a view to describing what customer-related proactiveness in developing radical innovations encompasses.

In order to analyze the change in customer-related proactiveness during *the process of radical innovation development*, studies on innovation development and radical innovations are combined with those concerning the market–technology linkage and analyses of specific marketing tactics used in certain innovation development phases.

Furthermore, international business studies, more specifically the ethnocentric, polycentric, regiocentric and geocentric (EPRG) model (see Heenan and Perlmutter 1979) that divides international orientation into ethno-, poly-, regio- and geocentric orientation, are used for analyzing the manifestation of the *international context* in proactiveness during the radical innovation development process. The EPRG model is apparently suitable since it is recommended that it be used in conjunction with an analysis of the firm's orientation toward its markets (Wind *et al.* 1973: 21). Finally, research on organizational features affecting the firm's behavior is utilized in order to describe *how customer-related proactiveness is created* in companies producing radical innovations.

All in all, it is worth emphasizing that the major theoretical background of this research lies in the above-mentioned theories that concentrate on customer-related proactiveness and the radical innovation development process. Thus, issues related to international business or fostering proactiveness in a firm are only considered on the level that is necessary in order to understand how they relate to the phenomenon in question, i.e. customer-related proactiveness in the development of radical innovations.

2 Proactiveness in the firm

The aim of this chapter is to clarify the concept of proactiveness and, in particular, to describe it in the context of the firm's behavior. It begins with a look at the firm–environment relationship in light of open systems and contingency theories. Reactive and proactive modes of behavior are then considered separately, with particular focus on proactiveness. However, as the two concepts form a continuous rather than dichotomous construct, the reactiveness–proactiveness continuum is presented next. The chapter ends with a description of a proactiveness pattern that is composed of the firm's positions along various reactiveness–proactiveness continuums.

2.1 A systems perspective on the firm's behavior

2.1.1 The firm as an open system

Systems theory is applicable to various fields. Its origins are in biology: the German biologist Ludwig von Bertalanffy presented the idea of general systems theory in a number of papers between 1945 and 1968. According to him, organisms (and also human organizations) are open systems. They are *systems* because they are composed of a number of interrelated subsystems, and they are *open* because feedback links connect them to their environments or the suprasystems of which they are a part (von Bertalanffy 1973; cf. Stacey 1993: 129).

Systems theory is holistic. It concentrates on the relations between systems on various levels and on their connections to the whole. The firm is seen as a system formed by various interrelated systems or subsystems. Its different functions do not behave in isolation but are interdependent. All parts influence all the other parts. Thus, every action has repercussions throughout the system. People occupy roles and conduct sets of activities in each system. They take part in forming and maintaining relationships with others both within their own system and in other systems or subsystems (Pfeffer and Salancik 1978; Scott *et al.* 1981: 44; Stacey 1993: 129).

There is a boundary separating each system from other systems and its environment, which could also be seen as a link between the system and the outside (Stacey 1993: 129, 134). According to Weick (1979: 132), this boundary may be rather unclear and, besides, depends largely on the person who is defining it.[1] According to the open-systems view, these boundaries are not closed. Each subsystem within a system, and each system within an environment, imports inputs (e.g. labor and materials) from other systems and exports outputs (e.g. products) to other systems. Thus, organizations are described as organic because they are seen to exhibit properties of life: i.e. they adapt to changing circumstances. They ingest resources, or inputs, from their environment and convert these inputs into outputs such as goods and services (Scott *et al.* 1981: 47).

Consequently, the firm and the environment are seen as being in a state of interaction and mutual dependence (Morgan 1997: 39–40; Stacey 1993: 130). The environment is seen as being composed of a complex composite of diverse factors (e.g. the product and labor markets, industry practices, governmental regulations), all of which influence the firm in their own particular ways (Miles and Snow 1978: 18). Environmental forces not only influence firms directly, they also interrelate with and affect each other, which creates secondary effects (Steele 1992: 25).

According to the systems perspective, complex interaction between variables makes the verifying of causal relations difficult, or even impossible. It is hard to distinguish between effects and causes, since the effect of a cause often becomes, in itself, the cause of a new effect (Brunsson 1989; Warren *et al.* 1998). Thus, the environment is not only seen as a given state to which firms adjust, it is also a dynamic outcome of their actions (Pfeffer and Salancik 1978: 190). Consequently, there is a dynamic interaction between a system and its environment, which includes the constant testing and comparing of discrepancies (Vancouver 1996: 172).

Open systems can prevent decay by acquiring energy from the environment and storing it in order to maintain their functioning. In practice, this means building up assets, which can be mobilized in times of recession or toughening competition, for instance (Khandwalla 1977: 226). Living open systems import information as well as energy. Negative information about their performance guides their next actions so as to enable the desired outcomes to be met. Hence, negative feedback helps the system to maintain its equilibrium and to learn from the consequences of its previous behavior[2] and provides the dynamics of its stability (Arbnor and Bjerke 1997: 122–124; Khandwalla 1977: 226; Stacey 1993: 117). However, although it solves current problems, it tends to ignore the reasons why these problems arose in the first place (Argyris 1994: 8–9; Hatch 1997: 371–372).

On the other hand, self-organizing systems can use both negative and positive feedback. Positive feedback makes the system change even more in the direction of deviation. Hence, negative feedback protects it from disturbances, but positive feedback enables it to broaden the deviations between its current situation and its goals, and to change its structure in order to cope better with the environment (Arbnor and Bjerke 1997: 122–124).

Using both negative and positive feedback loops to correct an organization's own behavior is called double-loop learning, and this enables the organization to define its own operating criteria (Hatch 1997: 372). It involves not only discovering what is changing in the environment and what consequences the firm's actions have, but also seeing what this all means for the unconscious models they are using. Double-loop learning is important because it can change mental models of acting and produce innovations (Stacey 1993: 181): the organization learns how to learn (Harryson 2002).

Living systems are thus self-regulating systems. The desired state can emerge from the system itself instead of from external agents (Vancouver 1996). Equifinality is a central concept within the open-systems view, meaning that there is no single way to achieve certain results: they may be attained in different ways, from different starting points and with different resources (Morgan 1997: 41). It thus enables choice in organizational decision-making (cf. Khandwalla 1977: 227).

Systems theory has traditionally emphasized the principles of equilibrium and homeostasis, and uncertainty has been regarded as a barrier to efficiency. Thus, in order to enable rationality[3] of operations at the core, it has been considered important to buffer core activities and to create bridging mechanisms to reduce the external turbulence (Hofer and Schendel 1978; Thompson 1974). However, lately the emphasis in systems theory literature has shifted to the analysis of instability and chaos (e.g. Boisot and Child 1999; Marion 1999; Pascale 1999; Warren *et al.* 1998). As shown in the following section, the research carried out by contingency theorists can be used for further analyzing the firm–environment relationship in conditions of both homeostasis and instability.

2.1.2 Contingency in the system

Contingency theory emerged during the 1960s. Inspired by Chandler's (1962) study about fit,[4] and building upon the legacy of the systems tradition (e.g. acknowledging equifinality and multiple goal seeking), it strongly influenced organizational analysis (Doz and Prahalad 1991: 151–152; Willmott 1990: 46). It encompasses the idea that there is a complex interrelationship between organizational variables and the conditions in the environment (Lawrence and Lorsch 1969: 47).

Traditionally, contingency theory has been considered rather deterministic. Causal connections between firms and the environment were regarded as one-directional. A certain environment was seen to cause a certain kind of successful strategy, and that caused a certain kind of successful structure (Chandler 1962; Stacey 1993: 61–62). Thus, it was concluded that contextual constraints have decisive effects on the operations of a firm (see Astley and Van de Ven 1983: 253; Chandler 1962). This deterministic view has been criticized by Bourgeois (1984), for example, who sees strategic management more as a creative activity in which free will also plays a role. However, it is worth noting that, even in the earliest writings on contingency theory, it was acknowledged that organizations were not solely determined by the environment but rather interacted with it (cf. Thompson 1974: 91–95).

An organization is considered to deal effectively with its environment when its internal states and processes are consistent with external demands[5] (Lawrence and Lorsch 1969: 47). The works of Chakravarthy (1986), Forte *et al.* (2000), Hambrick (1983), Hofer and Schendel (1978) and Miles and Snow (1978), for example, have supported the idea that a firm must find a "fit" between its environment, its strategy and its structure. Different environmental conditions require different organizing styles (Burns and Stalker 1979; Forte *et al.* 2000). Thus, the most effective solution is contingent upon the environment.

The fit between the firm and its environment determines the firm's performance (Lawrence and Lorsch 1969: 133–158). On the other hand, the level of performance also determines the range of strategic options available (Ginsberg and Venkatraman 1985: 423). Besides, it is not only the fit between the firm and its environment that is important, there also has to be a fit between the firm's subsystems for it to work effectively (Burrell and Morgan 1988: 154–159).

Although contingency theories generally regard strategy as a response to an exogenous variable, organizational conditions are also emphasized. It is acknowledged that there is no single strategy that is equally effective for all organizations. In order to match the firm's strategy to the dynamics of the environment, strategy managers need to possess good diagnostic skills so that they can understand the nature of the turbulent environment and thus develop strategies that best fit the environmental circumstances (see e.g. Abell 1978; Hofer and Schendel 1978; Luo 1999; Scott *et al.* 1981: 59). Nevertheless, contingency theory does not pay significant attention to the behavior of individual managers. Executive action is identified and measured only if it provides greater predictive certainty about the contingent relationship between the structural and the contextual variables (Child 1972: 16).

Doz and Prahalad (1991: 151–152) criticized contingency theory for failing to pay enough attention to the change in and adaptation to new environmental demands. However, it could be argued that, although studies applying the theory have traditionally concentrated on firm–environment alignment in static situations (e.g. Powell 1992; Venkatraman and Prescott 1990), the theory itself also allows research into dynamic change processes.[6] Since environmental conditions are in a continuous state of change, management must scan the environment repeatedly and make strategic decisions that are appropriate for the new external conditions (Itami and Roehl 1987; Lawrence and Lorsch 1969; Miles and Snow 1978; Thompson 1974). According to this view, strategy can be understood as a patterned stream of decisions concerning resource allocations in an attempt to reach a position consistent with the firm's environment (Mintzberg 1973: 53).

In general, later studies seemed to represent more proactive views on contingency theory. According to Itami and Roehl (1987: 9), strategic fit may be passive, active or leveraged. With a passive fit, the environment is taken as given, while an active fit implies that the firm can influence it. Leveraged fit means creating dynamic imbalance. This deliberate creation of a misfit between the firm's resources and its goals, thereby challenging it to accomplish the seemingly impossible, was suggested by Hamel and Prahalad (1989). They studied a number of global firms in America, Europe and Japan and found that the less-successful ones aimed to maintain a strategic fit with the environment, while the successful ones focused on using their resources in innovative ways in order to reach seemingly unattainable goals. Furthermore, they questioned the idea of adapting to the environment and proposed creative interaction as an alternative. Consequently, it seems that a firm can achieve a fit with its environment in both a reactive and a proactive way (cf. Boisot and Child 1999: 238).

2.2 The dichotomy of reactive and proactive behavior

It has been suggested by Løwendahl and Revang (1998: 769–770) that, given the complexity facing contemporary organizations, researchers could produce more managerial implications by looking for pragmatic concepts instead of causal relationships and explanatory models. Reactiveness and proactiveness are concepts that have been frequently used in the strategic literature in recent decades (e.g. Aragon-Correa 1998; Bateman and Crant 1999; Harper 2000; Larson *et al.* 1986). A large proportion of the literature on the strategic behavior of firms reflects a dichotomy between reactive and proactive strategic behavior (see e.g. Abell 1999; Tan and Litschert 1994).

2.2.1 Reactiveness

Reactive behavior has also been called "emerged" behavior (Vesalainen 1995: 43). Instead of formulating strategies for unforeseeable environmental changes, firms respond to the changes when they arise (Bennett 1996: 14). It has been claimed that planning for the future is not so useful in a turbulent business environment, and that firms have to live with uncertainty and chaos instead of looking for certainty where it cannot be seen: "All we can really do now is go with the flow and try to steer things a little" (Handy 1997: 22).

Since open systems are dependent on their environment, coping with environmental changes is a prerequisite for survival and prosperity. However, systems do not always even react to the changing environment, since inherent in them is a natural tendency to resist change and to continue as before (Arbnor and Bjerke 1997: 127–128). Environmental changes can be divided into the following three classes. First, *variation* means temporary deviation from the normal situation, to which the system usually reacts by altering its behavior within its existing areas of competence: since the environment will eventually return to normal, there is no need for fundamental changes. Second, *structural shift* is an irreversible change, to which systems have to adapt by changing their structures in order to maintain the fit. Third, *paradigmatic transformation* is a radical change in the environment, which demands a completely new approach (Arbnor and Bjerke 1997: 127–128).

According to systems theory, reactive behavior could be called error-controlled regulation and it is aimed at correcting deviations from the fit between

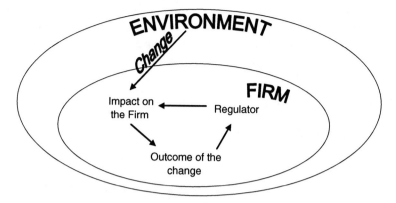

Figure 2.1 Reactive behavior in a firm.

the firm and the environment (Stacey 1993: 123). According to the principle of dynamic homeostasis, if the environment changes, the system must change too, but only to the extent that is necessary in order to maintain the fit with the environment (Khandwalla 1977: 226). This behavior is illustrated in Figure 2.1, which is based on Ashby's (1971: 226) model of error-controlled regulation in control systems (cf. Stacey 1993: 123).

According to this model, the regulator can sense the outcome of the environmental change once that outcome has occurred (Ashby 1971: 226). The regulator in this type of reactive behavior is the monitoring, reviewing and corrective action system of a firm, e.g. monthly meetings of the management in which they review what has happened to the business over the past month and consider how to correct deviations from the original plan (Stacey 1993: 123).

When faced with considerable changes, reactive firms struggle for survival,[7] but in a more slowly changing environment, they can manage by adjusting to the new situation (Abell 1999: 75–76). The problem with reactive behavior is usually the need for rapid action, in other words the lack of time and the few options to choose from (see de Geus 1988: 71). Thus, when such behavior becomes the principal tool in the firm's crisis management, there may be trouble ahead (see Bohn 2000).

Speed is not enough, however, and responses to changes in the environment ought to be appropriate. A system can adjust its strategic behavior either in a linear or a nonlinear manner (Warren *et al.* 1998). In the former, the response is proportional to its cause, while nonlinearity assumes a response that is more or less proportional. Nonlinearity occurs in both negative and positive feedback loops, and in positive loops in particular, it considerably increases the complexity of organizational behavior (Stacey 1993: 151–155).

Reacting to environmental changes is a complex process for most organizations, since it encompasses a vast amount of decision-making and behavioral choice on many levels (Miles and Snow 1978: 3). It is also difficult to analyze reactions to them because of the interacting set of nonlinear elements, which operate simultaneously and which are detected in and influenced by numerous feedback loops (Vancouver 1996). It is not sufficient to analyze reactive behavior if one wants to understand the firm–environment interface, however. It is equally important to take proactive behavior into account.

2.2.2 Proactiveness

Defining the concept of proactiveness

Analysis of the concept of proaction is rather thin on the ground. A [pro]active strategy has been defined by Jauch and Glueck (1988: 238) as "one in which strategists act before they are forced to react to environmental threats and opportunities." Thus, it seems that proaction is sometimes reaction: it is not reaction to the environmental change but rather to the symptoms of that coming change. This kind of behavior could be called anticipatory regulation (see Stacey 1993: 122–123), and its aim is to offset the undesirable effects of the disturbance on the outcome. In fact, *anticipation* could be understood as control of the present (because it is not possible to control the future) by using information about the past, present and future (Turtiainen 1999: 25).

However, proaction could also be seen as a firm's tendency to *influence* the environment and even initiate changes. Bateman and Crant (1993: 103–104) defined it as "a relatively stable tendency to effect environmental change" and "a behavior that directly alters environments." Correspondingly, Aragon-Correa (1998: 557) defined strategic proactivity as "a firm's tendency to initiate changes in its various strategic policies rather than to react to events." This reflects cognizance of firms' abilities to create their future, at least to some extent. For instance, they may imagine products, services and entire industries that do not yet exist and then give them birth (Hamel and Prahalad 1994: xi).

Consequently, the concept of proactiveness seems to include both anticipating and influencing. Proactive behavior is defined in this study in line with Johannessen *et al.*'s (1999: 118) definition: "Proactiveness is the ability to create opportunities or the ability to recognize or anticipate and act on opportunities (or dangers) when they present themselves." Accordingly, it is seen *to comprise both reaction to the symptoms of the coming change in the firm's environment and the creation of changes within it.* The separating factor between proactiveness and reactiveness seems to be time. Proaction always implies acting before the change in the environment has had a direct impact on the firm, and in practice it often involves reacting to the symptoms of that coming change (see Harper 2000: 80).

Some researchers (e.g. Perlmutter and Cnaan 1995) have equated proactiveness with entrepreneurship. However, the difference between these concepts seems to be that entrepreneurship is usually understood as discovering, evaluating and exploiting new opportunities (Shane and Venkataraman 2000: 218), whereas proactiveness includes the discovery, evaluation and management of

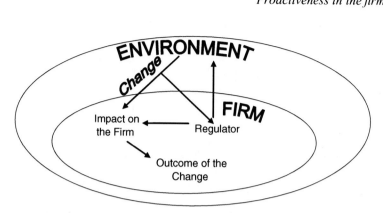

Figure 2.2 Proactive behavior in a firm.

environmental changes that could also present threats to the firm, not necessarily only opportunities.

Proactive behavior is illustrated in Figure 2.2, which is partly based on Ashby's (1971: 210) model of anticipatory regulation in control systems (cf. Stacey 1993: 123). This figure differs from Ashby's model with regard to the relationship between change and regulation. Ashby (1971: 210) saw this as a one-way directional relation from environmental change toward the regulator, which indicates that the regulator can sense the change before it hits the firm and take anticipatory action with a view to offsetting its undesirable effect on the outcome. However, given the definition of proactiveness applied in this study, the figure shows that the regulator can also influence the environment by creating changes in it in order to achieve the desired outcomes.

Although it has been suggested, that proactiveness is hard to observe because it is primarily a thought process (see e.g. March 1981; Weick 1988), there is a wealth of literature that mentions its characteristics. In order to be proactive, firms need to:

- actively collect information about non-customers and non-competitors and about probable changes in the environment (Drucker 1997: 22; Morgan 1992: 25);
- have industry foresight,[8] which is based on profound insight into future trends (in lifestyles, technology, demographics and geopolitics) that can be harnessed to rewrite industry rules and to create new competitive space (Hamel and Prahalad 1994: 76, 82);
- adopt an opportunity-seeking attitude, which implies developing and exploiting opportunities, being imaginative and creative concerning what the future could be, and utilizing this creative potential in shaping relationships with the environment (Hamel and Prahalad 1994: 82; Johannessen *et al.* 1999: 118; Morgan 1992: 26–27);
- strive to be leaders rather than followers (Morgan 1992: 28). Often this

means investing heavily in order to strengthen the firm's technological leadership (Dvir *et al.* 1993);

• adopt a "looking-ahead" approach and "driving in forward mode": the firm should shape rather than be shaped by change (Morgan 1992: 25–26);

• be capable of turning negatives into positives. Often this means tackling constraints (e.g. regulations or differing viewpoints) in a creative way (Morgan 1992: 27).

The different manifestations of proactiveness, i.e. anticipation and influence, are further described in the following sections.

Anticipating environmental circumstances

Anticipation requires detecting signals of change, following these signals and forecasting the coming changes, and acting accordingly (see Figure 2.3). Signals of change can be *detected* through environmental scanning, and macro-environmental scanning may help the firm to notice coming social, techno-logical, economic, environmental and political changes (Ashley and Morrison 1997: 48). The scanning may take place either at prespecified intervals or con-tinuously. On the one hand, it is not possible to anticipate all potential changes through periodic scanning, and on the other, undirected continuous scanning demands the unnecessary investment of resources. Therefore, it has been sug-gested that, in addition to periodic scanning, the firm should continuously monitor the potential changes in the areas it considers to be of most importance (Camillus and Datta 1991: 70). Naturally, the sooner it detects a coming change, the more action options it has (Ashley and Morrison 1997: 48; Camil-lus and Datta 1991: 70).

The environment can be divided into three levels depending on how it influ-ences a firm. The first level is the whole system of individuals, firms and other

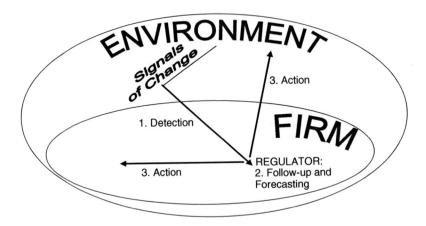

Figure 2.3 Anticipation of environmental change.

organizations that are interconnected to one another and to the focal firm. The second level comprises the individuals and organizations with which this firm interacts directly, and the third is the perception and representation of the environment (Pfeffer and Salancik 1978: 63).

Environmental analyses are also subject to various interpretations: managers respond not to the objective reality but to a set of their perceptions. Hence, different firms and even different individuals react differently to what seems to be the same environment (Pfeffer and Salancik 1978: 72–73; Smircich and Stubbart 1985: 724–727; Weick 1979; Weick 1988: 306–307). Furthermore, according to Weick (1979: 147–169), human actors do not react to an environment; they actually enact it. In other words, they influence the environment that influences them. Consequently, "what people refer to as their environment is generated by human actions and accompanying intellectual efforts to make sense out of these actions" (Smircich and Stubbart 1985: 726).

Attempting to grasp relevant information about coming changes and aiming to process this information rationally creates both ambiguity and uncertainty. Ambiguity results from an intersubjective process through which individuals in an organization attribute meaning to events and circumstances (Brownlie and Spender 1995: 40; Watkins and Bazerman 2003: 76–78). Uncertainty could be understood as an information defect (Spender 1993: 16). According to Spender (1989: 43–45), it affects the behavioral responses of a manager in various ways. First, the information may be incomplete, i.e. it may exist but is not at hand. Second, although the data may be complete, the phenomena cannot be grasped because the analyst is not capable of understanding the data. Third, although the data is available and the phenomena understood, the responses of the other actors cannot be known. Fourth, the analyst may be unable to process the knowledge in a rational manner (see also Brownlie and Spender 1995: 41; Spender 1993: 16–17). According to Brownlie and Spender (1995: 40), uncertainty can be resolved by managerial judgment. Although judgment can be learned from others, it mainly stems from experience and thus is highly contextually specific and difficult to change (Brownlie and Spender 1995: 45–48).

Scanned signals of change are *followed* through monitoring (Ashley and Morrison 1997: 48), the aim of which is to clarify whether they require action or not. Environmental issues that affect the firm in a way that is difficult to ascertain, i.e. "fuzzy issues," are particularly liable to create a crisis if they are ignored (Camillus and Datta 1991: 70). *Forecasting* helps to estimate the duration, direction and amplitude of the signals (Ashley and Morrison 1997: 48).

Organizational *actions* can be directed within the firm's internal or external environment (D'Aveni and MacMillan 1990). Internally directed actions help it to adapt to the environmental changes, whereas externally directed actions aim at modifying the environment (Chattopadhyay *et al.* 2001: 938). Firms often favor internally directed actions, since the managers find them easier to implement and control and less risky than externally directed actions (Chattopadhyay *et al.* 2001: 938; Cook *et al.* 1983: 203). Autogenic crises are examples of internally directed actions. These are crises that are deliberately initiated by the top management in order to prepare the firm for prospective crises (Barnett and Pratt 2000: 74–77). The following section focuses on externally directed actions.

Influencing environmental circumstances

It has been claimed that, in the long run, it is important for the firm to actively strive to create changes (Bennett 1996: 8; Handy 1997) and to discover strategic options of which the management was previously unaware (Wack 1992: 248). When the firm purposefully creates tensions in the environment, i.e. shakes the current fitness, it may be able to achieve even better fitness with the future environment (Beinhocker 1999; Pascale 1999). This kind of behavior, when instead of trying to tame the turbulence around it the firm absorbs it as an essential part of its strategic planning, has also been referred to as "riding the turbulence" (Boisot 1994: 46–47) (see Figure 2.4). Creating or precipitating environmental changes may stem from a desire either to surprise competitors or to change the course of action within a business, e.g. through setting new standards (cf. Evans 1991: 78–80).

Influence reflects power that is based on interdependence between groups and their contribution to common efforts (Stacey 1993: 42). Interdependence implies that each group is in a position in which it can, to some degree, facilitate or hinder the other's gratification (Emerson 1962: 32). Power could be defined as the ability of one group to get others to do something that they do not want to do, or could not otherwise do or would not otherwise have thought of doing (Benfari *et al.* 1986: 12; Pfeffer and Salancik 1978: 53–54; Stacey 1993: 41).

Kipnis *et al.* (1984) identified different dimensions of influence that managers use. Although their study concentrated on influence used within an organization, it also seems to be applicable to situations in which managers influence the environment of the organization (cf. Jokinen 2000). Thus, they may influence it through reasoning (i.e. using data and facts), friendliness (i.e. creating goodwill),

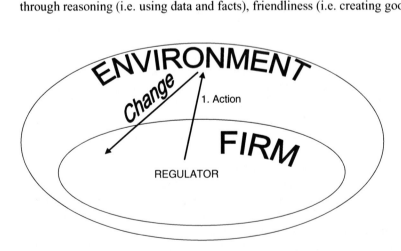

Figure 2.4 Influencing environmental change.

bargaining (i.e. exchanging benefits and favors) or assertiveness (i.e. using the direct and forceful approach). Furthermore, they may utilize coalitions (i.e. mobilize other actors), a higher authority (i.e. obtain support from a higher quarter) or sanctions (i.e. rewards and punishments) (Kipnis *et al.* 1984: 60–61).

The above-mentioned influence dimensions reflect explicit uses of influence. However, the environment may be also influenced implicitly: for example, by developing new products or opening new markets, a firm is able to create uncertainty in the competitors' environment and force them to react (cf. Miles and Snow 1978: 57).

This section has described ways in which the firm can act proactively and influence its environment. However, many writers have also argued that the firm's environment, especially in conditions of turbulence, may influence its degree of proactiveness. This is discussed in the following section.

Environmental turbulence influencing proactiveness

Consideration of how environmental turbulence influences the proactiveness of a firm should take into account the fact that environmentally driven changes in behavior are always based on organizational rather than objective perception of the environment (Pfeffer and Salancik 1978: 13). Therefore, if our aim is to understand the firm's behavior, our attention should be directed toward the environment as perceived by the people in it and not as perceived by the researcher, for example (Barr 1998; Khandwalla 1977: 342; Smircich and Stubbart 1985: 727).

The degree of environmental turbulence is an important feature of the firm's environment. Turbulence has been defined as "a measure of change that occurs in the factors or components of an organization's environment" (Smart and Vertinsky 1984: 200). A highly turbulent environment is a blend of significant growth opportunities and challenges to the firm's survival (Khandwalla 1977: 334). However, it is not the environmental turbulence itself that is problematic: it is a problem only when it involves factors that matter in terms of the firm's performance and survival (cf. Pfeffer and Salancik 1978: 68).

Distinguishing those key factors is nevertheless difficult, because they comprise both the factors that affect the task environment of the firm and those that operate in the broader contextual environment (Morgan 1989: 72). The concept of the "task environment" was introduced by Dill (1958: 410) to denote the forces that have an immediate influence on the firm's well-being (e.g. customers, suppliers, competitors and regulating groups). However, changes in the task environment often originate in the contextual environment (Morgan 1989: 72–73), and thus it would appear to be important to monitor and influence both.

According to Volberda (1998: 191–195), there are three basic dimensions of environmental turbulence: dynamism, complexity and unpredictability, all of which affect the turbulence simultaneously but not to the same extent (Figure 2.5). The unpredictability dimension is the most important contributor to environmental turbulence, and the complexity dimension is the least important.

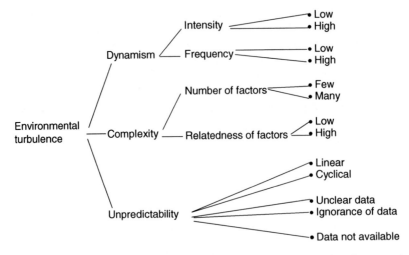

Figure 2.5 A taxonomy describing the dimensions and variables of environmental turbulence (adapted from Volberda 1998: 191).

The role of unpredictability is accentuated, since it often leads to a situation in which the firm is facing unfamiliar changes of which it has neither experience nor routine in terms of how to tackle them (Volberda 1998: 196–197).

Analysis of environmental *dynamism* requires distinction between the rate of environmental change (frequency) and its intensity. Volberda (1998: 192) emphasizes the importance of this division in stating that it is possible to undergo fast-occurring low-intensity changes (e.g. day-to-day fluctuations in demand) or substantial low-frequency changes (e.g. slow decreases in demand).

Complexity in the environment relates to the number of factors (including their diversity and heterogeneity) influencing the environment and their inter-relatedness (Child 1972: 3–4; Johnson and Scholes 1993: 77; Volberda 1998: 192–193). The bigger the degree of complexity, the more a plenitude of pertinent information is likely to be experienced by the management, and consequently the monitoring of diversified and heterogeneous information becomes more demanding (Child 1972: 3). Furthermore, the interrelatedness of the various factors further complicates the understanding of influence patterns (Johnson and Scholes 1993: 77).

Unpredictability in terms of the effects of the environmental forces may arise from their character. Linear and/or cyclical forces allow management to extrapolate past development, but when the relevant information is unclear, not available or ignored by the management, the unpredictability increases (Volberda 1998: 194–195).

It seems that, in the current business environment, dynamism and complexity in particular have generally increased. Changes are manifold and occur at an accelerating rate. Hence, the continuous management of change has become more important for the survival of firms, and it may even be necessary occasionally to create changes in the environment (Sauser and Sauser 2002: 34–36). Although this is difficult, and most turbulence is caused by exogenous factors beyond the control of a particular organization, a firm can act as a catalyst for environmental turbulence by bringing a radical innovation onto the market, for instance (D'Aveni 1999: 133).

There is *no consensus on whether proactiveness is best suited to a turbulent or a steady environment*. It has been argued that firms behaving proactively in a turbulent environment are better off than those who merely react (e.g. D'Aveni 1999: 128–130; Harper 2000; Slater 2001b: 170). However, the results of studies conducted by Kappel (2001), Liuhto (1999: 103–105), Smart and Vertinsky (1984), and Tan and Litschert (1994) indicate that, in fact, increased perceived turbulence is negatively related to proactive behavior and positively related to reactive behavior. Naturally, it is easier to behave proactively in a steady environment, whereas in a highly turbulent environment firms may be forced to react to unexpected changes.

2.3 The reactiveness–proactiveness continuum

According to Mintzberg (1994), effective strategies combine the ability to predict and the capability to react to unexpected circumstances. He called for a visionary approach in which the vision sets the outlines of a strategy but the details are worked out according to the circumstances (cf. Hansén 1991; Mintzberg 1991). Oktemgil and Greenley (1997: 462) and Liedtka (2000), to name but a few, have also advocated combining proactive and reactive behavior.

Many studies on reactiveness and proactiveness in firms have devised classifications based on their behavior. Many of these classifications (e.g. Chakravarthy 1982; Miles and Snow 1978; Vesalainen 1995) are derived from Simon's (1984: 12)[9] ideas that an open system can cope with the environment in three basic modes: passive insulation, reactive negative feedback and predictive adaptation.

One of the best-known studies on reactiveness–proactiveness was that conducted by Miles and Snow (1978). They presented a strategy typology and divided firms into four strategic archetypes: prospectors, analyzers, defenders and reactors. The basic assumption on which the typology is based is that firms act either to respond to the changing environmental context or to shape their environment, and prospectors (innovators) and defenders (firms that react slowly) hold extreme positions. Similarly, Hamel and Prahalad (1994: 28–29) divided firms on the road to the future into three classes: drivers (proactive firms), passengers (reactive firms) and road kills (very slow reactors or firms that do not even react at all). Harper (2000: 80) identified four types of firms: laggard firms, reactive firms, proactive firms and vanguard firms. The separating factor in this categorization is the time that it takes to adjust to the changes: whereas laggard firms need a major crisis to make them recognize the need for change, firms at the other extreme, i.e. vanguard firms, prepare themselves early.

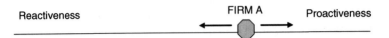

Figure 2.6 An example of a firm's position along the reactiveness–proactiveness continuum (Sandberg 2002: 186).

Even though proactiveness and reactiveness have been considered determinants of separate classes of firm behavior, some writers (e.g. Chakravarthy 1982;
Miles and Snow 1978) have emphasized the fact that the two concepts tend to
form a continuous rather than a dichotomous construct. Consequently, a firm
may adopt different degrees of proactiveness, find its place along the reactiveness–proactiveness continuum and, if necessary, also change its position relative
to either extreme (Figure 2.6).

This ability to change position is important in terms of the effective use of the
firm's resources. Proactiveness is expensive. Monitoring customers and competitors, long-range market scanning and extensive lobbying usually do not
come without a significant sacrifice of scarce resources. Therefore, it is natural
that if it were possible to wait and react, the firm would be tempted to favor
reactiveness.

Consideration of a firm's behavior as one position on a single proactiveness–
reactiveness continuum may lead to the oversimplification of such a complex
and multifaceted issue as organizational behavior, however. The possibility of
using several continuums is therefore discussed in the following section.

2.4 The proactiveness of the firm as a pattern

Larson *et al.* (1986) questioned the usefulness of the reactiveness–proactiveness
dichotomy in describing managerial behavior. According to their study, this
dichotomy, while useful in a general way for describing ideal managerial styles,
proved to be not as useful when applied to the behavior of particular managers.
In fact, the degree of proactiveness also seemed to be related to various task and
environmental factors (see also Luo 1999). Consequently, it could also be
argued that its mere position on the continuum may not adequately describe the
complex and multifaceted behavior of a particular firm, even though it may be
useful in a general way for comparing the behavior of many.

Hence, proactiveness can be analyzed separately in different functional areas
of the firm such as R&D, finance, operations, marketing, international and man-
agerial/structural organization (see Aaker and Mascarenhas 1984), R&D, production, marketing and finance (Chaganti and Sambharya 1987). Another option
is to distinguish between product-related, market-related and process-related
proactiveness (Hamel and Prahalad 1994; see Markides 1999: 3). Thus, there
appear to be many reactiveness–proactiveness continuums concerning products,
markets and processes, for example, and the firm has to position itself on each of
them.

The product dimension refers to reactiveness–proactiveness in product R&D and engineering and in manufacturing (cf. Srinivasan *et al.* 2002: 48–49; Urban and Hauser 1993: 33). Firms that behave reactively in this context produce imitations or modifications of products that are already on the market. On the other hand, firms that behave proactively introduce pioneering products and strive for first-mover advantages (cf. Simon *et al.* 2002: 187–190).

The market dimension encompasses the firm's market-related behavior. A firm that behaves reactively toward its market responds to the circumstances after having been influenced by them, while the proactive firm acts before the circumstances have had any effect. Since the market-dimension concept is an important one in this study, it is discussed in more detail in Section 3.1.

Processes refer to the way in which things are done (Sull 1999: 45) and include operations and distribution processes (Wright *et al.* 1995: 149) and management and administrative processes. Hamel (2000: 77) writes about core processes, which are methodologies and routines "used in translating competencies, assets, and other inputs into value for customers." Processes often become routines, which may attenuate the firm's response to environmental change (Sull 1999: 46–47). Thus, they may either enhance the firm's ability to influence or anticipate environmental change or, alternatively, they may be directed more toward reactive behavior.

Thus, *the degree of proactiveness in a firm could be seen as a pattern composed of its positions along these continuums*. For instance, Firm A in Figure 2.7 seems to be very proactive on both the product and the process dimensions, but in terms of the market it is still rather reactive.

The idea of contemplating a firm's behavior as a pattern is not new: it was suggested by Penrose (1980: 62)[10] and Cyert and March (1992: 102–103)[11] and supported by Miles and Snow (1978); for example, when trying to develop solutions to the problems of their firms, managers search only in the vicinity of familiar alternatives. In the course of time, this limited-search[12] activity becomes a routine and, in consequence, the firm may be well endowed in some areas

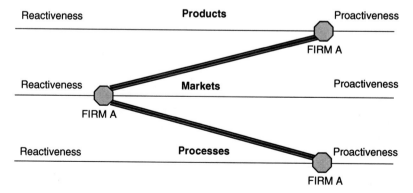

Figure 2.7 An example of a firm's positions on certain reactiveness–proactiveness continuums (Sandberg 2002: 187).

(e.g. R&D) but lack capabilities in others (e.g. marketing) (see Miles and Snow 1978: 8).

However, the pattern of proactiveness may also change if the circumstances so require and thus may be adapted according to the prevailing environmental or organizational contingencies (cf. Larson *et al.* 1986; Luo 1999). Adaptation can be defined as a firm's adjustment to a particular situation (i.e. certain circumstances) (Chakravarthy 1986: 438).[13] However, it is not enough in coping with continuously changing circumstances, and flexibility is also needed: flexibility is a dynamic concept that comprises the firm's ability to continuously adapt to changing circumstances (Aaker and Mascarenhas (1984: 74).[14]

According to Weick (1979: 135), adaptation in fact prevents flexibility, i.e. if an organization is well adapted in a specific environmental condition, it may be very difficult to adapt in changing circumstances. On the other hand, organizations designed for changing circumstances may have difficulties adapting well in any particular situation. According to McKee *et al.* (1989: 21), such firms are deliberately inefficient (cf. Brunsson 2000: 4). Hence, flexibility implies significant cost, and the level of flexibility varies according to the strategy of the firm (McKee *et al.* 1989: 21–22).

In sum, this section introduced a way of describing the proactiveness of a firm as a dynamic pattern, composed of its changing positions on different continuums. The next chapter concentrates on the market continuum.

3 Proactiveness toward customers

This chapter describes the firm's proactive behavior toward its customers. In order to understand proactiveness toward customers, it is first necessary to consider it in terms of the market in general. The focus thus turns toward customer-related proactiveness, which is one of the key concepts of the present study. The concept is first defined and described, and its international scope is discussed; the chapter ends with a look at the organizational factors that may foster proactive behavior toward customers.

3.1 Market proactiveness

Proactiveness toward the market can be understood as one aspect of market orientation, which in turn may be defined as "the organization wide generation, dissemination, and responsiveness to market intelligence"[1] (Kohli and Jaworski 1990: 3). Market intelligence covers the articulated and unarticulated, perceived and anticipated needs of current and prospective customers, as well as information on the exogenous market factors[2] that affect those needs (Culkin et al. 1999: 9; Kohli and Jaworski 1990: 3–4; Slater 2001b: 169–170). Consequently, the concept of "market orientation" incorporates both reactive and proactive behavior (cf. Berthon et al. 2004: 1067; Jaworski et al. 2000; Slater and Narver 1998: 1002–1003).

Market reactiveness means that the firm is market-driven:[3] it accepts the market structure as given and does not aim to change market behavior (Jaworski et al. 2000: 46–47). It can be measured in terms of the time it takes to respond to certain circumstances on the market after they have had a direct impact on the firm (see Harper 2000).

Consistently with the definition of proactiveness presented earlier (see Section 2.2.2), market proactiveness is defined here as either acting based on the information gathered about the market before the circumstances have had a direct impact on the firm or deliberately influencing and creating changes in the market.

Decisions on whether to approach markets proactively or reactively are made at the marketing-strategy level, and the context set for tactical marketing decisions, i.e. marketing-mix decisions (cf. Hultink et al. 1997: 245–246;

Zinkhan and Pereira 1994: 192–194). Thus, both strategic and tactical elements are likely to emerge when the market proactiveness of a firm is under consideration.

According to Brownlie and Spender (1995: 39), strategic marketing consists of future-oriented activities and decisions. Hence, the concept seems congruent with that of market proactiveness as defined above. However, a closer look reveals that it is often limited to the tasks of the marketing function (cf. Brownlie and Spender 1995; Cooper, L. 2000: 2–3; Slater and Narver 1998: 1003), whereas market proactiveness is the task of the whole firm. Further, the former seems to be limited to the anticipation of coming circumstances (see Brownlie and Spender 1995), and thus the question of influencing them is often passed over.

Anticipating the coming circumstances on the market implies that the firm can spot the first indicators of new market opportunities and risks and be among the first to exploit or prevent them (Hamel and Prahalad 1994: 83; Levitt 1960) (see Figure 3.1).

Influencing and creating changes in the market means that the firm shapes either the market structure or the behavior of the players[4] (Jaworski *et al.* 2000). The market structure incorporates a set of players (e.g. distribution channels, customers, competitors) and the roles they play. Jaworski *et al.*'s (2000) study on market influencing was basically restricted to the influence on demand and competition. This could be considered adequate, however, since influencing

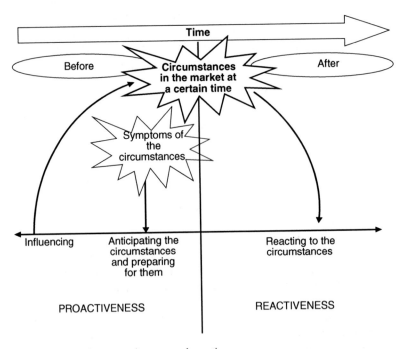

Figure 3.1 Market proactiveness and reactiveness.

other exogenous market factors such as government regulations could be seen in terms of influences on either competition (building/removing competitor constraints) or demand (building/removing demand constraints).

Thus, *market proactiveness is basically related to customers or to competitors.* There are two reasons why these two aspects should be dealt with separately (Li and Cavusgil 1999: 131–133): first, the cognitive activities are different (Day and Wensley 1988: 1; Li *et al.* 1999: 480), and second, depending on the situation, either the customers or the competitors can be emphasized in proactive actions (cf. Li and Cavusgil 1999: 131). Given the focus on customer-related proactiveness in this study, it is discussed in more detail in the following section.

3.2 Customer-related proactiveness

Defining customers is sometimes rather complicated. They may be end users or clients who influence or dictate the choices of end users, e.g. distributors (Kohli and Jaworski 1990. 4). Thus, the concept of the "customer" is extended to the end user's support group, management and distribution channel, for instance (Wah 1999: 19). Consequently, in business markets the customer may, in fact, be the user (who actually uses the product), the influencer (who influences the buying decision), the decider (who makes the buying decision), the approver (who authorizes the buying decision within the organization) or the buyer (the department that has the formal authority to buy and that acts as a gatekeeper in the organization) (cf. Kärkkäinen *et al.* 2001: 394; Rowley 1997: 83).

In terms of reactiveness and proactiveness toward customers, the key issue is whether customer behavior is reacted to, anticipated or influenced (see Figure 3.2). Given the definition of market proactiveness that was formulated in the previous section, *customer-related proactiveness could be defined either as acting based on the information gathered about the customers before their behavior has had a direct impact on the firm or as deliberately influencing and creating changes in behavior.* In this, customer needs seem to play a central role: in other words, they could be seen as the drivers of consumption or buying behavior (de Heer *et al.* 2002: 9).

Customer-needs can be defined as divergences between the existing and the desired situation (cf. Holt *et al.* 1984: 8; Kärkkäinen 2002: 6) and may exist or materialize in the future (Holt *et al.* 1984: 8). Existing needs can be further divided into articulated and latent needs. Latent needs are not apparent to customers, but they still exist and are unmet within the market (Jaworski *et al.* 2000: 51). Thus, they do not emerge onto the conscious level until the new product or service is presented (Holt 1976: 29). As long as these needs are not met, customers are satisfied because they are still ignorant of them (de Heer *et al.* 2002: 9).

Customer-related reactiveness implies that the firm concentrates on understanding and satisfying the articulated needs of its customers (cf. Slater and Narver 1998: 1001–1002). In practice, this is often accomplished through

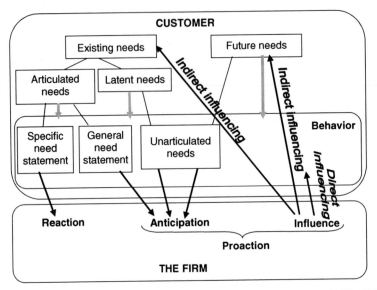

Figure 3.2 Customer needs and the firm's proactiveness (Sandberg 2007a: 255).

conducting market experiments, learning from the results and then modifying the offering based on this new knowledge (Slater 2001b: 170).

However, in cases in which the need statement of the customer is rather general (cf. Kärkkäinen 2002: 6), for example an articulation of problems with existing products or services, the firm may have to anticipate what kind of solution the customer really wants or needs. The anticipation seems to be particularly important in recognizing latent needs (cf. de Heer *et al.* 2002: 8–10; Slater and Narver 1998: 1001–1003), although it also seems applicable to the detection of future needs. As far as future needs are concerned, de Heer *et al.* (2002: 8) noted that, although the development of technology may be revolutionary, the needs of customers seem to develop in a rather evolutionary manner. Thus, it seems that anticipation is, in fact, possible.[5]

Customer behavior can be influenced directly, i.e. without regard to the cognitive structures, or indirectly, i.e. causing cognitive change, which then changes the behavior (Jaworski *et al.* 2000: 48). Constraints play an important role when it is a question of directly influencing customers. A firm may build customer constraints (real or imagined) that encourage purchasing by shaping expectations, for example, and it may also try to remove the constraints that prevent customer purchasing by shaping the choice sets, the criteria and the benefit packages they buy (Jaworski *et al.* 2000: 52).

Indirect influencing of customer behavior implies shaping their perceptions of offerings beforehand. On the one hand, a firm may create new needs by introducing a radical innovation into the market, informing its customers about the new benefits of the product and teaching them to use it.[6] On the other hand, it

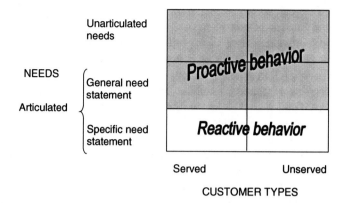

Figure 3.3 Proactive behavior and customer needs (adapted from Hamel and Prahalad 1994: 103).

may reverse existing customer preferences, thus affecting existing needs (cf. Jaworski *et al.* 2000: 51–53).

Figure 3.3 illustrates the relationship between proactive behavior and customer needs. It is based on the classification put forward by Hamel and Prahalad (1994: 100–105), who utilized a two-by-two matrix to show how well a firm meets the articulated and unarticulated needs of both served and unserved customers. In order to make the matrix applicable to the description of proactive behavior, articulated needs are further divided into specific and general need statements. Reactive behavior means that the firm reacts to specific need statements of customers, whereas the proactive firm meets either their unarticulated needs or their general need statements. Whether the customers are existing (served) or new (un-served) seems insignificant in the analysis of proactiveness toward them.

Thus, customer-related proactiveness seems to be manifested in various ways. Its analysis is even more multifaceted, however, when the international dimension is taken into account. This is discussed in the following section.

3.3 The international scope of customer-related proactiveness

Although past research has provided an extensive picture of the innovation development process, the geographical linkage has received rather limited attention, despite some emphasis on the role of location and proximity (e.g. Gertler 1995; Gerybadze and Reger 1999; Porter and Sölvell 1999). Porter and Sölvell (1999) identified three different characteristics of the process, which accentuate the importance of the geographical dimension. First, there is a need for the incremental reduction of technical and economic uncertainty, which requires continuous interaction with the firm's environment; second, there is also a need to get ideas, information and resources from outside the firm; and third, there is a need for face-to-face contacts in particular, in order to improve communication in the exchange and creation of new knowledge.

Global and local influences tend to vary along the innovation development process (Lindqvist *et al.* 2000). As far as customer-related proactiveness is concerned, it should be taken into account that proactiveness is expensive. Monitoring customers, long-range market scanning and extensive lobbying are not normally possible without a significant sacrifice of scarce resources (cf. Quinn 1979: 24–28), and the costs are therefore more justifiable if targeted on the appropriate geographical area.

Classic ethnocentric, polycentric, regiocentric and geocentric (EPRG) typology is used in this study in order to analyze the international scope of action. According to Heenan and Perlmutter (1979), the international orientation of the firm can be classified into four types: ethnocentric (E), polycentric (P), regiocentric (R) and geocentric (G). Although the typology was originally developed and is often used (e.g. Gassmann and von Zedtwitz 1999; Schuh 2001) to describe the development of a multinational organization, it also seems to be suitable for assessing the way in which a firm perceives international markets in general (see e.g. Shoham *et al.* 1995; Wind *et al.* 1973). It emphasizes the geographic dimension and seems to be suitable for comparing behavior related to a particular product, whereas other internationalization models such as that of international marketing evolution developed by Douglas and Craig (1989), and Johanson and Vahlne's (1977) internationalization process, would be more suitable when focusing on the firm and its internationalization path. The EPRG typology also seems to allow the assessment of international orientation during the whole innovation development process, whereas approaches focusing on market entry (e.g. Ayal and Zif 1979; Hollensen 2001: 203–206; Keegan 1969) concentrate mainly on the launch stage. It also seems to be rather compatible with analyses of market proactiveness, since it has been argued that consideration of both EPRG orientation and the degree of market orientation[7] is needed in order to understand international market strategies (Wind *et al.* 1973: 21).

Ethnocentric orientation means that the focus is on the home market (Heenan and Perlmutter 1979: 18–19). Consequently, ethnocentric customer-related proactiveness can be defined as the process of influencing and anticipating the coming market circumstances based only on information gathered from customers in the home country (cf. Wind *et al.* 1973: 15). Thus, the product strategy is based on the home-market customers' needs, while marketing strategies and tactics are usually transferred from the home market to other markets (Shoham *et al.* 1995). The weakness in the ethnocentric orientation is the lack of sensitivity to signals from foreign markets, which leads to insufficient consideration of different local-market demands (Gassmann and von Zedtwitz 1999: 236).

The emphasis in *polycentric orientation* is on the differences between countries (Heenan and Perlmutter 1979: 18–19). Consequently, polycentric customer-related proactiveness implies that the firm influences and anticipates customers' behavior in various countries individually (cf. Wind *et al.* 1973: 15). The product strategy is adapted to customers' needs in the host market, and market strategies are adapted to each market (Shoham *et al.* 1995)[8]. Markets that attract attention are usually selected based on the degree of familiarity or knowledge (Douglas and Craig 1989: 53). Although polycentric orientation is optimal for each local market, it may also be very costly and lead to inefficiencies (Gassmann and von Zedtwitz 1999: 240–241).

Regiocentric orientation accentuates specific orientation toward a regional grouping of countries (Heenan and Perlmutter 1979: 18–19). Thus, regiocentric customer-related proactiveness is anchored on a certain region, and the product strategy is based on the needs of customers within it (Shoham *et al.* 1995: 11). It has been suggested that the current trend toward regionalism is strengthening this orientation. Hence, firms are benefiting from the harmonization of standards and regulations, as well as from the convergence of customers' tastes within regions. Nonetheless, the trade barriers at the outer borders and the differences in customer tastes worldwide constrain the regional benefits (Proff 2002: 243–244).

Geocentric orientation implies a global approach (Heenan and Perlmutter 1979: 18–19). In terms of customer-related proactiveness, this means that the firm is trying to influence and anticipate the behavior of customers on a global scale. Consequently, the product strategy is standardized[9] and based on global needs (Shoham *et al.* 1995: 11). It has been claimed that customers around the world will have more similar needs in the future as globalization spreads (Wah 1999: 21). In particular, needs concerning product technologies seem to be converging (Hu and Griffith 1997: 117), which may create more avenues for geocentric orientation. However, the danger is that critical local market requirements may be ignored (Gassmann and von Zedtwitz 1999: 239).

Consequently, *the international scope of customer-related proactiveness may vary, ranging from the ethnocentric to the geocentric.* However, directing proactiveness toward international markets is rather demanding, as the geographic distance increases the expense of anticipating or influencing customer behavior (Li *et al.* 1999: 477–478). Furthermore, proactiveness is also more difficult when it extends to different countries: for example, data problems (Hutcheson 1984, 53) and difficulties associated with cross-cultural research (Parameswaran and Yaprak 1987; Yu *et al.* 1993) tend to complicate matters with international customers.

Market knowledge of foreign markets can be classified into experiential and objective: the former is acquired from the firm's experiences of direct interaction in the market, whereas the latter can be obtained through training or the acquisition of data (Johanson and Vahlne 1977: 28; Penrose 1980: 53). Li *et al.* (1999: 481) claim that experiential knowledge is more useful in designing new products for foreign markets, whereas objective knowledge tends to be helpful especially in the early phases of internationalization. Thus, experiential knowledge could be thought of as playing a key role in developing radical innovations for international markets.

In sum, this section dealt with the international scope of customer-related proactiveness, whereas the previous section described the concept. Since it was argued that both the degree of proactiveness and its international scope may vary, it is interesting to consider how proactiveness can be promoted in a firm. This is dealt with in the following section.

3.4 Fostering customer-related proactiveness in the firm

As was stated in Section 2.1.2, according to contingency theory, organizational variables influence the firm's behavior. The variables that, according to previous literature, seem to play a major role in creating proactiveness in general and customer-related proactiveness in particular are briefly described in this section.

3.4.1 Organizational structure

Organizational structure comprises the distribution of responsibilities and authority among the personnel, the planning and control systems, and the process regulations concerning decision-making, coordination and execution (cf. Mintzberg 1979: 2; Volberda 1996: 364). Hence, the structure could be understood as facilitating the interaction and communication in terms of organizational behavior. The structure of a firm is often analyzed by contemplating its position along the mechanistic–organic continuum (see Table 3.1) (Burns and Stalker 1979: 119–125; Khandwalla 1977: 411; Slevin and Covin 1990: 44).

Generally, organic firms are more open, more loosely controlled and more informal (Slevin and Covin 1990: 43–44; Volberda 1996: 364). There tends to be a high degree of participation in decision-making, a low degree of hierarchical authority (El Louadi 1998: 182) and a tendency to work in multidisciplinary teams (Mintzberg 1979: 435). Mechanistic firms, on the other hand, are more constrained, more hierarchical and more formal (Slevin and Covin 1990: 43–44;

Table 3.1 The continuum of organizational structures

	Mechanistic organizational structure	*Organic organizational structure*
Communication	Restricted	Open
Managerial style	Uniform	Allowed to vary freely
Decision-making	Superiors make decisions with minimum consultation and involvement of subordinates	Participation and group consensus used frequently
Adaptation	Reluctant	Free
Emphasis	Formally-laid-down procedures	Getting things done
Control	Tight	Informal
On-job behavior	Constrained	Flexible

Source: Adapted from Khandwalla (1977: 411); Slevin and Covin (1990: 44).

Volberda 1996: 364), which means that their behavior is predictable and standardized (Mintzberg 1979: 86).

According to contingency theory, there is no one best structure, and the appropriateness of a certain type of structure depends on the environmental circumstances (e.g. Khandwalla 1977: 412; Miles and Snow 1978: 251–254). Firms with a mechanistic structure are adapted to stable conditions, whereas those with an organic structure are adapted to changing conditions (Burns 1963; Burns and Stalker 1979).

Covin and Slevin's (1988) study suggests that, in general, an organic structure best supports proactive behavior. Enhanced communication and minimized bureaucratic barriers foster initiating actions and enable rapid response (Slevin and Covin 1990: 43–44). Kohli and Jaworski (1990: 11) propose that the rigidity inherent in a bureaucracy apparently decreases market orientation. Furthermore, Slater and Narver (1995) have argued that the anticipation and creation of customer needs are characteristics of a learning organization and that organizational learning is facilitated through an organic structure. Consequently, since an organic structure implies open communication and cross-functional cooperation, it might be presumed that it furthers customer-related proactiveness throughout the firm (cf. Barrett and Weinstein 1998: 60). Thus, we could conclude that an *organic structure apparently contributes to customer-related proactiveness.*

3.4.2 Leadership and management

Both leadership and management seem to influence proactive behavior in the firm. *Leading* can be defined as influencing and guiding (in direction, action and opinion) (Darling and Nurmi 1996: 261). It involves setting a direction for the firm, communicating this direction to the employees, and motivating and inspiring them to move in the right direction (Kotter 1990: 104–109). On the other hand, *managing* is more concentrated on accomplishment and driving (Darling and Nurmi 1996: 261). It encompasses activities such as planning, budgeting, organizing, staffing, controlling and problem solving (Kotter 1990: 104–109). According to Kotter (1990: 103), leadership complements management and both are needed for success. Both types of activities have been analyzed in light of their contribution to the firm's proactiveness.

Bateman and Crant (1999) described different ways in which managers can foster proactive behavior in the organization. First, individuals with a proactive disposition can be identified and selected from among job applicants. Second, proactive behavior of employees can be enhanced via training by emphasizing skills such as problem finding and creativity. Third, by liberating and relaxing organizational formalities (i.e. fostering an organic structure), managers can create more space for proactive behavior. Fourth, inspiring people to act proactively includes highlighting the importance of proactive behavior in achieving the organizational mission, encouraging people to have new ideas and tolerating failure (Bateman and Crant 1999: 66–67).

Nadler and Tushman (1990) studied the behavior through which a leader can initiate, lead and manage strategic reorientation in an organization. Reorientation was understood as proactive change management, which makes this study interesting as far as the focal study is concerned. They identified patterns of behavior that are needed to manage both the people and the organizational context in the firm in order to make it behave proactively in changing circumstances.

These authors referred to managing people as charismatic leadership, which consists of three basic components: envisioning, energizing and enabling. Envisioning means creating a vision with which people can identify and which can inspire them. Energizing implies motivating people, e.g. by expressing confidence and demonstrating the leader's own enthusiasm. Enabling involves helping people to accomplish their tasks in the face of challenging goals and is thus more related to emotional assistance, e.g. empathizing and supporting (Nadler and Tushman 1990: 80–83).

Managing the organizational context is called instrumental leadership in the same study. It involves three elements: structuring, controlling and rewarding. Structuring means creating structures that support the behavior that is required throughout the firm (Nadler and Tushman 1990: 85); controlling mainly involves the creation of systems and processes for measuring, monitoring or assessing behavior and results (Lawler and Rhode 1976: 2–6); and rewarding involves the administration of rewards and punishments in order to enhance the desired behavior (Nadler and Tushman 1990: 85).

Hansén (1990) offered another perspective on the role of management. He referred to creative leadership that aims to foster all-round creative thinking, which is needed in the firm for pioneering and managing constant change. Although he did not explicitly bring up the term proactiveness in his article, it could be considered an implicit manifestation of creative thinking. In other words, creative thinking means pioneering and doing things differently (Hansén 1990: 56, 58).

The elements involved in creative leadership are direction, knowledge, framework and follow-up. The creative use and combination of these elements enable management to bring out the creativity of employees, which is likely to increase enthusiasm and commitment throughout the organization. *Direction* encompasses vision, strategy and implementation. It is a continuous process of strategic thinking, which evinces both constant curiosity toward changing circumstances and a desire to search for new opportunities and challenges (Hansén 1990: 60). The components of *knowing* are structure, order and relation. In order to expand creativity, it is useful to consider each component individually, and to understand how problem solving is often restricted if they are used in a rather limited, routine way (see Hamel 2000: 117–144; Nierenberg 1982: 39). Fostering cooperation within the organization management contributes to creative knowledge management. However, what often seems to be forgotten is that a good atmosphere is a prerequisite for the creation of cooperation. The main areas in the *framework* are organization, technology and culture, and it is the task of creative leadership to encourage flexibility within it. For example, the framework needed for idea generation differs considerably from one that would maximally support the commercialization of the invention. *Follow-up* should first and foremost focus on time, money and communication: it is crucial for organization-wide creative thinking that employees understand follow-up as open monitoring based on constructive discussion (Hansén 1990: 62–65).

The aspects mentioned by Bateman and Crant (1999) seem to be rather similar to the characteristics of the leadership of change mentioned by Nadler and Tushman (1990) and to the features of creative leadership described by Hansén (1990). However, whereas Bateman and Crant (1999) concentrate more on management, the aspect of leadership seems to be more strongly present in the research conducted by Nadler and Tushman (1990) and Hansén (1990). The separate classifications are presented together in Figure 3.4.

It thus seems that, through leadership and management, it is possible to foster the proactive behavior of a firm in various ways. The different text frames in Figure 3.4 are used to indicate the building up of proactive capabilities (hexagon), enabling (retangle) and motivating (oval) proactive behavior.

If the above-presented studies on the role of managers in increasing proactiveness are considered together with studies on the role of managers in

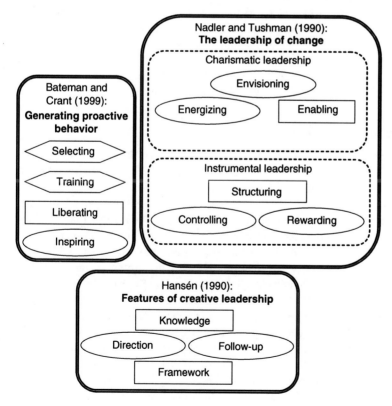

Figure 3.4 Aspects of leadership and management fostering proactiveness in a firm.

enhancing market-orientation, it is possible to draw certain conclusions concerning their role in fostering customer-related proactiveness. First, it seems that customer-related proactive capabilities can *be built up* by selecting people who have both a proactive disposition and knowledge/concern about customer needs and by further training employees in the skills that enhance both customer focus and creative problem solving (cf. Bateman and Crant 1999: 66–67; Harris and Piercy 1997: 35–37).

Second, customer-related proactive behavior can *be enabled* in a firm. This may include offering emotional support for this kind of behavior, or creating structures that facilitate it (cf. Nadler and Tushman 1990: 80–85). Organizational structures that permit cooperation and open communication within the firm have been considered important in fostering proactiveness in general (cf. Hansén 1990). Moreover, it appears that structures that enhance interfunctional cooperation (cf. Jaworski and Kohli 1993: 64) and open vertical communication between employees at the customer interface and head-office managers (cf. Harris and Piercy 1999: 124–125) increase customer-related proactiveness.

Third, it is apparently also possible *to motivate* customer-related proactive behavior by continually reminding employees how important it is to anticipate and influence customer behavior (cf. Bateman and Crant 1999: 66–67; Jaworski and Kohli 1993: 64), which suggests that it is worthwhile monitoring customer-related behavior and rewarding employees for this. Furthermore, taking evolving customer needs into account in planning may enhance customer-related proactiveness throughout the firm (cf. Harris and Piercy 1997: 36–37; Jaworski and Kohli 1993: 64–65; Nadler and Tushman 1990: 80–85). This brings us to the subject of organizational culture, which is discussed next.

3.4.3 Supportive organizational culture

Since the organizational culture affects the behavior in a firm (Ahmed 1998: 33; Alvesson and Berg 1992: 28), it is also likely to affect its proactiveness (cf. Kotter and Heskett 1992: 43–46). Organizational culture can be defined as a shared system of meanings that distinguishes one organization from another and conveys significance to events and circumstances (cf. Shrivastava 1985: 103). It can also be understood as the generally accepted way of solving problems in the organization (Ahmed 1998: 32). Although culture may be rather stable, it is never entirely static (Kotter and Heskett 1992: 7). Furthermore, it can be at least partly managed, controlled and changed (Alvesson and Berg 1992: 29).

Homburg and Pflesser (2000: 450–451) studied various definitions of organizational culture and came to the conclusion that market-oriented culture consists of shared basic values, behavioral norms, artifacts and behaviors. They also found that market-oriented shared basic values enhanced market-oriented norms, and that these norms, in turn, influenced market-oriented behavior through certain artifacts (Homburg and Pflesser 2000: 457). Thus, it seems that *shared basic values, behavioral norms, and artifacts that emphasize the role of customers may also foster customer-related proactive behavior in the firm.*

Organization-wide shared basic *values* reflect what organizational members consider desirable (Shrivastava 1985: 105) and important (Ott 1989: 39). They are conscious and emotion-laden (Ott 1989: 39). Values are used in decision-making and in the evaluation of outcomes (Brown 1992: 4), and hence they govern and shape the behavior in a firm (cf. Kotter and Heskett 1992: 5; Schein 1984: 3). Certain values may be disseminated and furthered by recruitment and promotion policies that favor those with the "correct" values (Alvesson and Berg 1992: 83).

From previous studies, it could be assumed that values that tend to increase customer-related proactiveness in a firm include the following:

- long-term commitment to understanding customers' latent and expressed needs (Slater and Narver 1998: 1002);
- the profitable creation and maintenance of superior customer value without neglecting the interests of other key stakeholders (Slater and Narver 1995: 67);
- an outward focus and propensity for action (Webster 1994: 14);
- risk-taking (Jaworski and Kohli 1993: 64; Kremer and O'Brien 1994: 43);
- openness of internal communication (Homburg and Pflesser 2000: 450–451; Webster 1993);
- responsibility among employees (cf. Calori and Sarnin 1991: 59–71; Homburg and Pflesser 2000: 450–451).

Norms are shared expectations regarding what are appropriate or inappropriate attitudes and behaviors (O'Reilly 1989: 12). They are formed based on basic values, and they legitimate specific behaviors (cf. O'Reilly 1989: 13; Shrivastava 1985: 105). They have two dimensions: intensity and crystallization. Intensity reflects the amount of approval attached to the certain attitude or behavior, whereas crystallization refers to how widely the norm is shared throughout the organization. Both of these dimensions are needed for creating a strong, consistent culture (O'Reilly 1989: 13; cf. Kotter and Heskett 1992: 15–18).

Specific norms related to the array of values presented above include the following:

- customer focus;
- the openness of customer-related internal communication;
- the propensity to act toward customers;
- the willingness to accept occasional market failures;
- the customer-related responsibility of employees (cf. Homburg and Pflesser 2000: 450–451; Jaworski and Kohli 1993: 63; Slater and Narver 1995: 67).

Artifacts are material and non-material things that communicate information about the organization's values and behavior. Material artifacts include annual reports, brochures, and physical layouts and arrangements of the premises, while organizational language, stories, myths, ceremonies and celebrations are examples of non-material artifacts (cf. Alvesson and Berg 1992: 80–85; Dandridge *et al.* 1980; Ott 1989: 24–35; Trice and Beyer 1984). Although artifacts are the most visible layer of the organizational culture, they are hard to interpret (Schein 1984: 3).

Artifacts can be also used to enhance customer-related proactiveness. For example, according to Webster (1994: 14), details of language and other symbols may capture and communicate the vision of customer orientation. Likewise, Ahmed (1998: 38) argues that corporate statements can be used to guide behavior and to express the organizational culture, for example. Furthermore, as Homburg and Pflesser (2000: 451) state, it can be assumed that stories about ideal proactive customer-oriented behavior, an open and friendly customer entrance and reception area, regular awards for employees behaving proactively toward customers and a customer-focused discussion style during meetings are some of the things that tend to further enhance such behavior in the firm.

The organizational culture is closely linked to the individuals working in the organization, and they are discussed in the next section.

3.4.4 Individuals predisposed to customer-related proactive behavior

The proactive behavior of an organization is grounded in the behavior of the individuals working in it (Bateman and Crant 1993: 105): without their behavior there would be no corporate behavior (Rollinson 2002: 16; Weick 1979: 34). Furthermore, since people's characteristics have a bearing on their behavior (Rollinson 2002: 67), these characteristics may also be reflected in the behavior of the firm (cf. Holt *et al.* 1984: 32).

Studies on proactive individuals are based on the tradition of trait studies, which presume that enduring traits are consistent across situations and predictive (cf. Bem and Allen 1974: 506–507). However, it is worth emphasizing that if someone has a particular trait, it only indicates that he or she has a predisposition to behave in predictable ways in particular circumstances. Hence, traits are not regarded as inerrable guides to individuals' actions (Rollinson 2002: 85).

Traits can be further divided into basic and surface traits. Basic traits (e.g. agreeability) are dispositions to behave in diverse situational contexts, whereas surface traits are dispositions to behave in specific situations. Thus, both customer orientation and customer-related proactiveness can be understood as surface traits. (cf. Brown *et al.* 2002: 111–112) (see Figure 3.5).

The study by Brown *et al.* (2002) suggests that both agreeability and the need for activity are the kind of basic traits that increase customer orientation, i.e. both proactive and reactive behavior toward customers. Agreeability means compassionate interpersonal orientation (Bateman and Crant 1993: 106), and it can be assumed that employees high in agreeability may feel empathy with the customers and be more willing to solve their problems. Furthermore, such people may also derive personal satisfaction from solving customers' problems and helping them to satisfy their needs (Brown *et al.* 2002: 112; Hogan *et al.* 1984; Hurley 1998: 119).

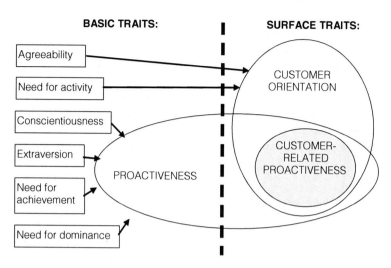

Figure 3.5 Basic traits related to customer-related proactiveness.

The need for activity is manifested in the selection, evocation and manipulation of features of the environment (Buss 1991: 480–481). Individuals possessing a high need for activity tend to complete more tasks and tend to be more active in everyday life. It seems that the desire to keep busy and to stay active is an important predictor of customer orientation. Employees with a low need for activity are less likely to be motivated to work at meeting customer needs, especially in contexts that require a degree of activity (cf. Brown *et al.* 2002: 113).

Bateman and Crant (1993) studied personal disposition to proactive behavior. An individual with a prototypic proactive personality has been described as someone who is relatively unconstrained by situational forces and who influences environmental change (Bateman and Crant 1993: 105). More specifically, proactive employees have been characterized as task- and job-competent, interpersonally effective, organizationally oriented, committed to united goals, initiators, outspoken and oriented to higher values (Campbell 2000: 55–57). According to Bateman and Crant's study (1999: 64), business people predisposed to proactive behavior tend to scan for change opportunities, to set effective change-oriented goals, to anticipate and prevent problems, to do different things or do things differently, to take action, to persevere and to achieve results.

Bateman and Crant (1993) suggest that proactive behavior is related to conscientiousness, extraversion, the need for achievement and the need for dominance. *Conscientiousness* corresponds to the degree of organization, persistence and motivation in goal-oriented behavior and seems to be related to proactiveness, since both are goal-oriented behaviors that accentuate persistence. *Extraversion* has been described as a need for stimulation, assertiveness, activity and quantity and intensity of interpersonal interaction. It implies seeking new experiences and activities, just like proactiveness. Furthermore, both *the need for achievement* and *the need for dominance* are traits that influence the environment and thus seem to share some construct overlap with proactiveness (Bateman and Crant 1993: 106–107).

In sum, although there are no studies concerning the basic traits that are directly related to customer-related proactiveness, those describing the relations between basic traits and both customer orientation and proactiveness could be considered indicative. Hence, it is assumed that *individuals predisposed to customer-related proactiveness are also likely to possess traits such as agreeability, conscientiousness and extraversion and the need for activity, achievement and dominance.*

3.4.5 Liquid financial assets

It is not only human but also material resources that influence the customer-related proactiveness of a firm (cf. Chakravarthy 1982: 37). However, previous literature seems to lack consensus regarding the impact of material resources on customer orientation per se. Studies conducted by Slater and Narver (1998) and Sinkula *et al.* (1997) indicate that an abundance of resources increases customer orientation. However, discovering customers' latent needs is demanding and may require more sophisticated market-research techniques (Slater and Narver 1998: 1003), thus the costs of acquiring that knowledge may rise. Furthermore, allocating resources to the creation of customer intelligence usually does not produce an immediate payoff: on the contrary, the payoff tends to be uncertain and long term. In any case, the generation of customer intelligence requires continuous investment since it can quickly go out of date (cf. Sinkula *et al.* 1997: 307; Slater and Narver 2000: 125). Harris and Piercy (1997) questioned the idea that customer orientation necessarily requires additional costs, however. They argued that mere reorientation of existing resources may suffice.

Although there is some contradiction regarding the impact of resources on customer orientation, researchers seem to agree that *resources enhance proactive behavior.* According to Quinn (1979: 24–28), firms that lack resources tend to behave reactively, while those with abundant resources are more likely to employ proactive strategies. Slack resources in particular have often been seen as important facilitators of proactive behavior (Bourgeois 1981: 29; Floyd and Lane 2000: 154). Bourgeois (1981: 30) defined organizational slack as:

> that cushion of actual or potential resources which allows an organization to adapt successfully to internal pressures for adjustment or to external pressures for change in policy, as well as to initiate changes in strategy with respect to the external environment.

Thus, slack resources allow the firm to interact with its environment more boldly (Bourgeois 1981: 35).

Even though Miles (1982: 248–251) divided organizational slack into economic (liquid financial assets), political (e.g. social commitments and goodwill) and managerial slack (management capacity), slack is normally identifiable in monetary terms, i.e. in *liquid financial assets* (cf. Chakravarthy 1982: 42). According to Chakravarthy (1986: 450), indicators of slack financial assets are profitability, productivity and the ability to raise long-term resources.

In practice, evaluating the firm's slack resources is difficult, since managers are not willing to reveal the details in interviews (Bourgeois 1981: 32). Thus, they have often been measured from public financial records (Chakravarthy 1986). There also seem to be certain difficulties in using financial records, however, since they represent a snapshot of a certain moment rather than the whole picture of the firm's affairs over a longer period (Bourgeois 1981: 37). Besides, public financial records may give a rather erroneous picture of the firm's actual slack, particularly of those producing radical innovations. These innovations are often born in small firms, where the founder-inventor may invest a large amount of his or her own money, and where additional funding may emerge from heterogeneous sources, thus making it difficult to track down from the financial statements (Allen 2003: 189–229).

The connection between slack resources and the firm's previous performance was made by Cyert and March (1992: 189),[10] who stated that "success tends to breed slack." However, this is not always the case, since profits can be paid back to the shareholders[11] instead of being used to build up the material and organizational capacities that help the firm to better adjust to the changing environment (Chakravarthy 1982: 42).

In sum, it appears from previous studies that liquid financial assets increase the proactiveness of the firm and therefore may be assumed also to foster proactive behavior toward customers.

3.4.6 Complementary external linkages

On the above evidence, it seems that an organic structure, facilitative leadership, a supportive organizational culture, proactive individuals and liquid financial assets tend to increase the firm's customer-related proactiveness. All these contributors could be considered resources, i.e. "firm-specific assets that are difficult if not impossible to imitate" (Teece *et al.* 1997: 516). Assets may be tangible or intangible (Teece 2000: 15–16), and they may or may not appear in the balance sheet (cf. Teece *et al.* 1997: 521–522). Due to their scarcity and the increasing concentration on core competencies, firms are forced to acquire some of their assets from outside (cf. Wigand *et al.* 1997: 210). A large array of different types of firms and organizations can contribute to the firm's innovation activities (Gemünden *et al.* 1996: 450), and assets can be acquired from them either through single transactions or through longer-term cooperation.

Cooperation can be classified, in terms of the direction of collaboration,[12] as vertical, horizontal or diagonal. Vertical cooperation takes place inside the value chain (e.g. between manufacturer and retailer), horizontal cooperation exists between similar firms operating in similar markets, and diagonal cooperation occurs between businesses from different branches (Wigand *et al.* 1997: 224).

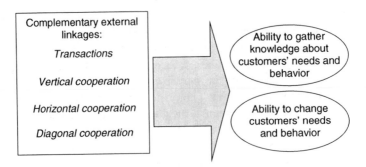

Figure 3.6 External linkages contributing to the firm's customer-related proactiveness.

Collaboration tends to reflect more than just a formal contractual exchange,
especially in high-tech industries. Numerous handshake deals and informal
forms of collaboration usually accompany and substitute formal agreements
(Powell *et al.* 1996: 120).

Given the above discussion on customer-related proactiveness, we could say
that a firm can foster its proactiveness *through its external linkages* in two basic
ways: first, *it can acquire more knowledge of customer needs and behavior, and
second, it can jointly influence both the needs and the behavior of customers* (cf.
Slater and Narver 2000: 122) (see Figure 3.6).

According to Ritter and Gemünden's (2003) study, the ability to initiate,
handle and utilize a portfolio of interorganizational relationships increases
innovation success. Using the collaboration to generate intelligence about customers' needs and behavior seems to be particularly important in situations in
which the knowledge base of an industry is complex and evolving. In these
cases, collaboration may reduce the uncertainties associated with radical innovations and new markets (Powell *et al.* 1996: 116–118; Slater and Narver 2000:
122). A common assault on customer needs and behavior is particularly important in situations in which the new innovation has to function with other products
or when compatibility is needed in creating the new standard and, consequently,
assuring fast market acceptance (Hillebrand 1996: 385).

Thus, a firm may increase both its knowledge about customers and its influence over them through complementary external linkages, which thus may be
utilized to foster proactiveness toward them.

3.4.7 Synthesis

In sum, organizational structure, leadership and management, organizational
culture, the characteristics of individuals working in the firm, financial assets
and external linkages all potentially contribute to the customer-related proactiveness (Figure 3.7).

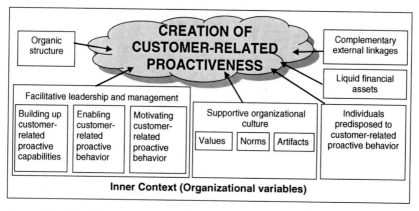

Figure 3.7 Organizational variables influencing the creation of customer-related proactiveness.

On the basis of previous studies, it was further concluded that an organic organizational structure in particular seems to foster customer-related proactiveness. Furthermore, managers can build up customer-related proactive capabilities and enable and motivate customer-related proactive behavior. Similarly, proactiveness can be valued, expected and manifested in the organizational culture through its values, norms and artifacts. It also seems to be important to have individuals predisposed to customer-related proactive behavior working in the firm. Moreover, since proactiveness demands resources, slack resources, and more specifically liquid financial assets, may also foster the firm's customer-related proactiveness. Finally, further fostering is possible through complementary external linkages, which can contribute to knowledge creation and influence power.

4 Developing radical innovations

In order to understand the development of radical innovations, it is first necessary to clarify the concept of radical innovation and discuss its general features. The radical innovation development process and different ways of describing it are then discussed. Given the results of previous studies, the development process is divided here into three phases: idea generation, development and launch. Customer-related proactiveness and its manifestations in each stage are then discussed and, finally, a framework for describing the firm's proactive behavior toward its customers during the whole innovation development process is presented.

4.1 The concept of radical innovation

In order to understand the concept of radical innovation, it is first necessary to define the concepts "idea," "invention" and "innovation." New ideas are the starting point for an innovation. According to Trott (2002: 12), an idea can be a concept, a thought or a collection of thoughts. As long as customers regard the idea as new, it is an innovative idea, even though it may not be "objectively" new. Hence, the key issue in determining the novelty of an idea is customer perception (Robertson 1971: 6; Rogers 1983: 11). Sometimes the new idea stems from a *discovery*, i.e. the recognition of previously unknown natural laws (Kuhn 1970: 52–53). An idea becomes an invention when it is converted into a tangible artifact (Trott 2002: 12) (see Figure 4.1).

According to Schumpeter (1934: 88–89), innovation is not synonymous with invention; whereas inventions are economically irrelevant, innovations include the idea of economic leadership or commercial success (see also Schumpeter 1934: 128–156). This division was later clarified in terms of the different basic objectives. Rogers (1983: 138) put forward a process perspective, according to which invention is discovering and creating a new idea, whereas innovation means its adoption or use. A more teleological view is evident in the division suggested by Hansén and Wakonen (1997: 346), who state that whereas invention aims at solving a technological or scientific problem, innovation solves a commercial problem. This emphasis on economic success can also be seen in other definitions (see e.g. Cumming 1998: 22; *Innovation in Small and Medium Firms* 1982: 21–22; Rickards 1985: 10–11). Thus, *in order to become an*

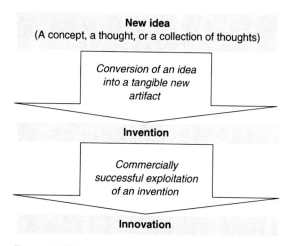

Figure 4.1 The path from an idea to an innovation.

innovation an invention has to be commercially successful, which presumes both a successful product or service launch and diffusion in the market. Consequently, one does not necessarily lead to the other (Schumpeter 1927: 293).

Innovations can be classified into product, process, organizational, management, production, commercial and service innovations (Trott 2002: 13–14). This study concentrates on those that are developed and sold to outside customers, i.e. product and service innovations. Another common classification method is to distinguish between novel and incremental innovations (e.g. Ali 1994; de Brentani 2001; Song and Montoya-Weiss 1998). Very novel innovations have been called by various names in previous literature (Table 4.1).

It is clear from the above table that, on the one hand, different terms have been used for a concept that has been defined in similar ways and, on the other hand, that different definitions have been used for the same term. The definitions given can be analyzed in terms of their *perspective*, i.e. for whom the innovation is novel. Some of them (e.g. Anderson and Tushman 1990: 607; Veryzer 1998a: 307) seem to take the firm's perspective and concentrate on how much the innovation differs from those already on the market (cf. Atuahene-Gima 1996: 94), whereas others (e.g. Moore 1999b: 13; Veryzer 1998a: 307) see the newness from the customer's perspective, i.e. how compatible the innovation is with the experiences and consumption patterns of customers (cf. Atuahene-Gima 1996: 94). Nevertheless, most (e.g. Chandy and Tellis 1998: 475; de Moerloose 2000) seem to include both the firm and the customer perspectives, and a few (e.g. Trott 2002: 236; Urban *et al.* 1996: 47) even add a macroperspective and take into account the impact on prevailing market structures.

The definitions could also be analyzed in terms of *scope*, i.e. what in the innovation is novel. These novelty aspects seem to include technology (e.g. Ali 1994: 48), benefits (e.g. O'Connor 1998: 152), behavior (e.g. LaPlaca and Punj

Table 4.1 A review of definitions of novel innovations

Author(s)	Term	Definition of the term[a]
Ali (1994: 48)	Pioneering products	"Technological breakthroughs"
Anderson and Tushman (1990: 607)	Product discontinuities	"...fundamentally different product forms that command a decisive cost, performance, or quality advantage over prior product forms"
Chandy and Tellis (1998: 475; 2000: 2)	Radical product innovation	A new product that incorporates a substantially different technology from existing products and can fulfill key customer needs better than existing products
de Moerloose (2000: 75–77)	Breakthrough products	Products that are technologically new (e.g. include new raw material, a new production process, technical change, input replacement), commercially new (i.e. create originality by commercialization), functionally new (i.e. provide new benefits/new solutions for the customers), and induce new customer groups
Hansén (1981: 56)	Breakthrough discoveries	"...products which cover a completely new principle of treatment" (a study was conducted in the pharmaceuticals industry)
LaPlaca and Punj (1989: 92)	Discontinuous innovations	"...entirely new products or services that require or cause a drastic modification of the users' purchasing or consumption behavior, such as enabling the user to perform an entirely new function"
Leifer et al. (2001: 102)	Radical innovation	"...a product, process, or service with either unprecedented performance features or familiar features that offer significant improvements in performance or cost that transform existing markets or create new ones"
Moore (1999b: 13)	Truly discontinuous innovations	"...new products or services that require the end user and the marketplace to dramatically change their past behavior, with the promise of gaining equally dramatic new benefits."
New Products ... (1982: 8–9)	New-to-the-world products	Products that are new to the firm and that create an entirely new market
O'Connor (1998: 152)	Discontinuous innovation	The creation of a new product or process that has either unprecedented performance features or familiar features that offer the potential for fivefold to tenfold improvements in performance or cost

O'Connor and Rice (2001: 99)	The creation of a new product or process that has either unprecedented performance features or familiar features that offer the potential for fivefold to tenfold (or greater) improvements in performance, or a 30–50% (or greater) reduction in cost
Radical innovation or Breakthrough innovation	
Robertson (1971: 7)	Innovation that "involves the establishment of new consumption patterns and the creation of previously unknown products"
Discontinuous innovation	
Trott (2002: 236)	Products that differ from existing products in that field, sometimes creating an entirely new market, and then they require buyers to change their behavior patterns
Discontinuous new products	
Urban et al. (1996: 47)	Products that "shift market structures, represent new technologies, require consumer learning, and induce behaviour changes"
Really-new products	
Veryzer (1998a: 307)	Products involving significant new technologies regardless of whether or not they are perceived by the customers as really new
Technologically discontinuous products	
Veryzer (1998a: 307)	"…products that are perceived by customers as being really new regardless of whether or not they utilize new technology"
Commercially discontinuous products	

Note
a The definitions in quotation marks are presented by the author(s) as definitions of the concept. Others have not been formally expressed by the authors but are interpreted as being in the nature of definitions and are therefore included in this review.

1989: 92), markets (e.g. *New Products Management for the 1980s* 1982: 8–9) and commercialization (e.g. de Moerloose 2000: 75–77).

The terminology regarding novel innovations is thus rather confusing. The term "radical innovation" is used consistently in this study to refer to novel innovations. Since the study concentrates on customer-related proactiveness in developing radical innovations, it is important to come up with a definition that particularly reflects the customer perspective (cf. Allen 2003: 265). The newness of technology is also included, although customers may not necessarily even be aware of it. However, since technological novelty often plays an important role in customer evaluation and adoption of innovations (Veryzer 1998b: 138), it is included in the definition. Thus, the term is defined as *a new product or service, which requires considerable change in customer behavior, is perceived as offering substantially enhanced benefits and is also technologically new.* Features of radical innovations are discussed in more detail in the following section.

4.2 Characteristics of radical innovations

As stated above, radical innovations require considerable changes in customer *behavior*. They may establish new consumption patterns (Robertson 1971: 7) or change established ones (Lawton and Parasuraman 1980: 20–21). If the innovation is perceived as difficult to use and understand, i.e. complex, it is not usually adopted easily. On the other hand, the ability to test it enhances its adoption, since trial potential reduces customer uncertainty (Rogers 1983: 230–231). The role of trials in innovation adoption was stressed by Robertson (1971: 240–241), who argued that through dissonance reduction[1] trials may reinforce an initially favorable attitude toward the innovation or even turn negative attitudes positive.

It is worth pointing out that, although radical innovations create changes in customer behavior, this does not mean that they necessarily create changes in consumer needs. For example, when digital cameras and microwave ovens were first introduced, there already existed a prevailing product that satisfied customers' needs for capturing memories and cooking food (Pinkerton 2001: 117).

For the firm, the ultimate goal of radical innovation is to create revenue and corporate wealth (Kaplan 1999: 16). In order to create revenue and succeed in the marketplace, the innovation needs to provide new *benefits* for customers. By pursuing radical innovations, firms may replace prevalent methods of delivering value to customers (Kaplan 1999: 17) and provide them with a completely new level of functionality (O'Connor 1998: 152; cf. Mohr 2001: 16). However, the benefits perceived by customers seem to be more important than the objective benefits (cf. Veryzer 1998b: 138). Customers usually evaluate these benefits by contemplating the degree to which the innovation appears to be better than the product it displaces. In other words, customers evaluate its relative advantage. The relative advantage is often expressed in terms of economics or status, and since it incorporates the perceived benefit or reward of adoption, it is positively related to the rate of adoption (Rogers 1983: 213–217).

However, since the benefits are not known, they need to be conveyed to the potential customers (Guiltinan 1999: 521). The more observable they are, the faster the innovation tends to be adopted (Rogers 1983: 232). Providing information about benefits is particularly difficult in a situation in which the product is still non-existent, and potential customers are forced to visualize its functionality and benefits based merely on concept descriptions. This is often the case with radical innovations (cf. Mullins and Sutherland 1998: 228).

According to Veryzer's (1998b) study, the key factors that influence customers' evaluations of radical new products differ from those that are important in evaluating incremental products. First, the lack of familiarity may cause resistance, and even fear, among customers. Second, the newness of these products may encourage customers to focus on irrational product attributes that may not correspond to their real requirements. Third, user–product interaction problems require customers to invest time and effort in learning to use the product properly. Fourth, customer uncertainty about the benefits of these products, not to mention the risks associated with them (this often arises from the lack of other comparable products), encourages resistance. Fifth, aesthetic aspects affect customer feelings about product safety and their attraction to it: aesthetic aspects therefore seem to play an important role even during the customer testing of early prototypes. Sixth, the collision between these products and customers' life and consumption patterns causes resistance (Veryzer 1998b: 144–146). The benefits of an innovation tend to be more easily observable when they are compatible, i.e. consistent with existing values and beliefs, with previously introduced ideas and with the needs of potential adopters (Rogers 1983: 223–226).

Radical innovations also often expand *technological* capabilities (cf. Ali 1994: 48; Tushman and Anderson 1986; Veryzer 1998b: 138). Technological novelty in this context may mean technical change in the product features or the use of new raw material, for example (cf. de Moerloose 2000: 75). However, significant technological novelties, or disruptive technologies,[2] are not necessarily a prerequisite of radical innovation. For example, the SONY Walkman could be regarded as radical in its effect on behavior and in its benefits, although it was not extremely novel in terms of technological capability. On the other hand, technological novelties alone may not lead to radical innovation. For instance, the switch in televisions from vacuum tubes to solid-state technology only created incremental changes in customer behavior and benefits (cf. Veryzer 1998b: 138). These kinds of situations, in which opinions concerning the novelty of an innovation differ between the innovating firm and the consumers, also pose considerable challenges in terms of marketing (cf. Mohr 2001: 16).

Radical innovations may cause extensive changes in the innovative firm's environment when they appear: i.e. they often create their own *market* (Guiltinan 1999: 511; Mohr 2001: 16; Nayak and Ketteringham 1986: 341; O'Connor 1998: 152; Shanklin and Ryans 1984: 165). They give the firm a possibility to create new competitive space (Kaplan 1999: 17) and shift market structures (Urban *et al.* 1996: 47), and sometimes (e.g. in the cases of the first PCs and automobiles) they define a new industry (Veryzer 1998b: 138). Furthermore, often a whole new infrastructure has to be created in order to deliver the full benefit (Jolly 1997: 11).

Radical innovations are characteristically new both to the firm and to the outside world (Kleinschmidt and Cooper 1991; *New Products Management for the 1980s* 1982: 8–9; Olson *et al.* 1995: 52). According to Veryzer (1998a: 306), these innovations "involve dramatic departures from existent products or their logical extensions." Therefore, they often also face resistance inside the firm. Individual decision makers may find it very difficult to risk supporting a development project that might eventually fail because there is no market for the product (Christensen 1997: 160).

O'Connor (1998: 152) argues that the challenges inherent in the management of radical innovation have been largely disregarded by academics. Radical innovations, indeed, seem to involve unique *challenges in terms of both development and commercialization* (cf. Ali 1994; Mohr 2001: 18): in the former because of the high level of uncertainty concerning their technological feasibility, and in the latter because of the uncertainty involved (Veryzer 1998b: 138). If there is a high level of uncertainty, innovation development practices should have the potential to reduce risk and manage uncertainty while concurrently enhancing the likelihood of success in the market. However, since it is difficult for potential customers to articulate the needs that a radical innovation may fulfill, it is hard and expensive to identify market opportunities[3] and to translate technological advancements into products that meet customer needs (Mullins and Sutherland 1998). Furthermore, since the development of these products may span 10–20 years, or even more, the future needs of customers may differ considerably from their current ones (Christensen 1997: 208; Veryzer 1998a: 318).

Shanklin and Ryans (1984: 165) argue that radical innovations are often developed in supply-side markets, where buyers' desires are rather presumed than identified (cf. Christensen 1997: 147–150; Millier 1999: 15). However, this contradicts the results of the studies by von Hippel (1986, 1989, 2006) and von Hippel *et al.* (2000) that emphasize the role of lead users in initiating ideas for new radical innovations. Therefore, it could be concluded that radical innovation is a two-sided phenomenon that often incorporates both the internal push of new technological advances and the external pull of market need to ensure the necessary eventual commercial success.

Nevertheless, despite the risks inherent in developing radical innovations, they may bring considerable *rewards* in the future (through competitive advantage and differentiation, for example) (Song and Montoya-Weiss 1998: 125). The relationship between innovativeness and success is not necessarily straightforward, however. According to Kleinschmidt and Cooper (1991), there is a U-shaped relationship between innovativeness and certain success measures. This is a consequence of the enhanced opportunities for differentiation and competitive advantage, on the one hand, and of the increased costs and uncertainty involved in the development of the radical innovation, on the other.

When a firm brings a new radical innovation onto the market, it has a temporary monopoly position. Later on, as the volume of production and demand grows and a larger variety of applications is introduced, the competition gets harder (Utterbach and Suárez 1993: 2). This diffusion of proprietary technology can be delayed by patent protection (Porter 1980: 172). In fact, according to Acs *et al.* (2001: 240), "Innovations arise only when property rights are properly protected." However, Porter (1980: 173–174) considers this patent protection to be quite unreliable, because patents may be evaded by new inventions. He argues that it would be more sustainable to create new technology continuously through research and development.

In sum, in order to achieve success, an innovation has to be developed to meet market needs. This seems to be a rather demanding task, however, due to its radical features. The challenges inherent in the development process are discussed further in the following sections.

4.3 Features of the radical innovation development process

New product development has traditionally been seen as a process of uncertainty reduction (Urban and Hauser 1993: 17). This is rather difficult for radical innovations in that considerable new knowledge and information to reduce uncertainties are needed at a time when the ultimate uses for the radical ideas or inventions are often still unknown. In these cases, the learning has often been called "probing and learning," which signifies that the development team is learning during the development process (Christensen 1997: 46, 99; Lynn *et al.* 1996).

Thus, the process of radical innovation entails the firm and the customers creating and discovering the new market together. This often renders the analytical and decision-making processes used for continuous innovations (e.g. economic modeling, trend analysis, interviewing leading customers...) rather superfluous: on the contrary, action often has to be taken before careful plans can be made (Christensen 1997: 147–150, 160; Millier 1999: 50–56).

This does not necessarily obviate the need for planning but rather changes the focus from implementation toward learning, i.e. the planning becomes discovery driven (Christensen 1997: 160). Discovery-driven planning acknowledges the need to envision the unknown. It is recognized that at the start of the process "little is known and much is assumed." Thus, underlying assumptions are taken as best-guess estimates to be questioned. In the course of time, newly uncovered data is incorporated into the changing plan, and the real potential of the new product is discovered along with its development (McGrath and MacMillan 1995: 44–45). Christensen (1997: 161–162) calls the discovering of emerging markets for radical technologies "agnostic marketing," which implies that there is no knowledge about the final use of the radical product at the beginning of the development process.

The discovery of emerging markets, together with the uncertainties concerning new technologies, also shapes the process (Veryzer 1998b). Various researchers have presented their own models, most of which are linear and define the various stages, although it has been frequently stated that, in practice, these stages are overlapping rather than sequential and do not always proceed in the same order (Rogers 1983: 149; Trommsdorff 1995: 4). It has also been argued that new-product development, in fact, tends to be more linear than is usually assumed (Wind and Mahajan 1997: 7).

An example of the structured and linear process model is the stage-gate™ approach presented by Cooper (1988), which breaks the innovation development into clearly predetermined stages, each one consisting of certain prescribed activities. The key idea in this line of thought is that each stage is a gate at which the decision whether to continue or not is made. Stages and gates are seen to help in managing risk by requiring certain key questions to be addressed in order to proceed further.

The cyclical model of innovation management presented by Van de Ven *et al.* (1999) gives a very different view of innovation development. This model describes it as a non-linear cycle of divergent and contingent activities (e.g. creating ideas, executing relationships), which may be repeated over time, take place at different organizational levels, and depend on various constraining and enabling factors. The process is not formalized or highly structured. For example, clear product and opportunity definition can often be built up only after prototype development. Moreover, the phases are overlapping rather than discrete, and coincidence and serendipity have an important role (Jolly 1997: xix–xx, 41–42; Veryzer 1998a: 313–318).

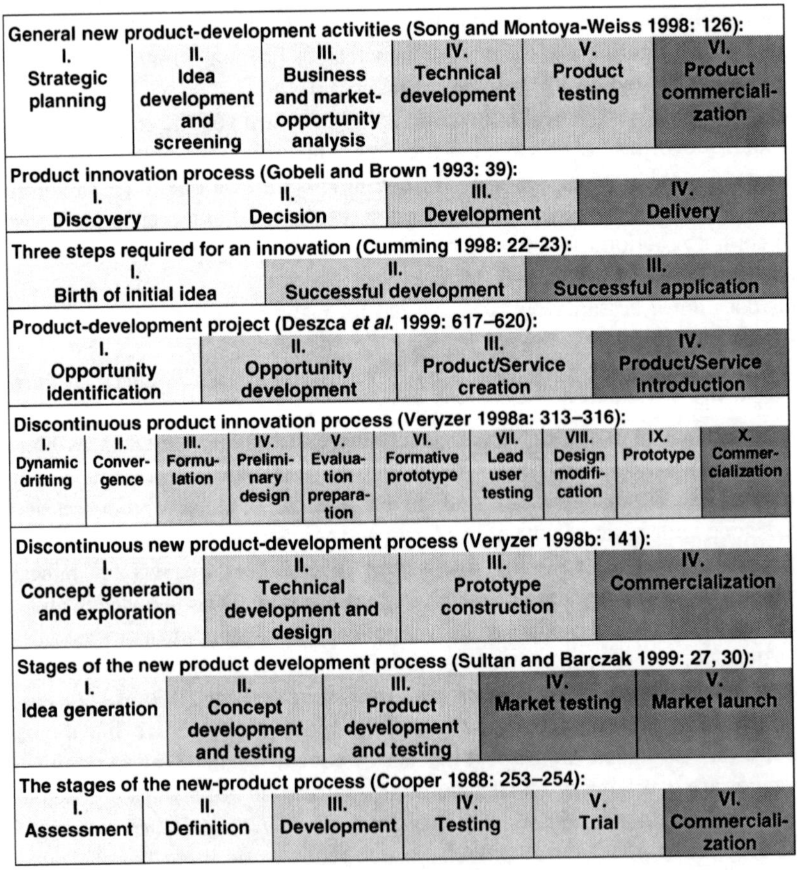

General new product-development activities (Song and Montoya-Weiss 1998: 126):

I. Strategic planning	II. Idea development and screening	III. Business and market-opportunity analysis	IV. Technical development	V. Product testing	VI. Product commercialization

Product innovation process (Gobeli and Brown 1993: 39):

I. Discovery	II. Decision	III. Development	IV. Delivery

Three steps required for an innovation (Cumming 1998: 22–23):

I. Birth of initial idea	II. Successful development	III. Successful application

Product-development project (Deszca *et al.* 1999: 617–620):

I. Opportunity identification	II. Opportunity development	III. Product/Service creation	IV. Product/Service introduction

Discontinuous product innovation process (Veryzer 1998a: 313–316):

I. Dynamic drifting	II. Convergence	III. Formulation	IV. Preliminary design	V. Evaluation preparation	VI. Formative prototype	VII. Lead user testing	VIII. Design modification	IX. Prototype	X. Commercialization

Discontinuous new product-development process (Veryzer 1998b: 141):

I. Concept generation and exploration	II. Technical development and design	III. Prototype construction	IV. Commercialization

Stages of the new product development process (Sultan and Barczak 1999: 27, 30):

I. Idea generation	II. Concept development and testing	III. Product development and testing	IV. Market testing	V. Market launch

The stages of the new-product process (Cooper 1988: 253–254):

I. Assessment	II. Definition	III. Development	IV. Testing	V. Trial	VI. Commercialization

Figure 4.2 The process of innovation development.

The stages of new-product development (Sommers 1982: 54–56):					
I. Ideation	II. Screening	III. Business analysis	IV. Preliminary development	V. Advanced development	VI. Commerciali-zation

The life cycle of technological innovations (Millier 1999: 29–31):		
I. Research project life cycle	II. Technological innovation life cycle	III. Product life cycle

The key sub-processes in bringing new technologies to market (Jolly 1997: 3–12):				
I. Imaging	II. Incubating	III. Demonstrating	IV. Promoting	V. Sustaining

High-technology product-development process (LaPlaca and Punj 1989: 99):		
I. Assessment	II. Development	III. Execution

Innovation development process (Rogers 1983: 135–149):					
I. Recognition of a problem or need	II. Basic and applied research	III. Development	IV. Commer-cialization	V. Diffusion and adoption	VI. Consequences

New product-development model (Trott 2002: 212):							
I. Idea generation	II. Idea screening	III. Concept testing	IV. Business analysis	V. Product develop-ment	VI. Test marketing	VII. Commer-cialization	VIII. Monitor-ing and evalua-tion

New product-development process (Urban and Hauser 1993: 38):				
I. Opportunity identification	II. Design	III. Testing	IV. Introduction	V. Life-cycle management

Product life cycle stages (Baker and Hart 1999: 96–101):						
I. Gestation	II. Introduction	III. Growth	IV. Maturity	V. Saturation	VI. Decline	VII. Elimination

Innovation development process (basis of the framework of this study):		
I. Idea Generation	II. Development	III. Launch

Figure 4.2 continued.

A linear approach is used in this study because it is more simple, it is extensively used and it seems to provide a sufficient basis for anchoring this rather multifaceted phenomenon: customer-related proactiveness. This approach is supported by Gobeli and Brown (1993: 39), for example, who suggest that clustering the innovation development activities into stages seems to give a common and practical format for discussion of the problems and solutions during the process. Further support is to be found in the study conducted by Veryzer (1998a). He acknowledges that the development of radical innovations differs considerably from that of conventional new products: "[Radical] innovation seems to be an inherently messy process" (Veryzer 1998a: 318). However, he argues that, even though the process is less structured, there is a logical progression in how it is managed. Thus, *activities occur in a consistent sequence* even in the development of radical innovations (Veryzer 1998a: 313).

Separate models of the innovation development process are presented, and eventually synthesized, in Figure 4.2. As can be seen, even though various process models for developing radical innovations, or new products, have been suggested, the basic progression of activities is rather similar.

Most of the models illustrated in the figure begin from the idea-generation phase and end at the stage in which the product has been commercialized. Accordingly, in this study too, the process of innovation development is understood to extend from idea generation to development stage[4] and, finally, to the launch stage. Although it would be interesting to analyze customer-related proactiveness during the whole life cycle,[5] it seemed more important to concentrate more thoroughly on the first stages, which have been claimed to be the most determinant for the success of an innovation (e.g. Bacon *et al.* 1994: 40–45; Khurana and Rosenthal 1998).

The idea-generation stage is marked in light grey in Figure 4.2, and the same color is used to indicate the corresponding phases in the previous models. Similarly, the development stage is marked in darker grey and the launch stage in black. These different stages are discussed in detail and combined with the analysis of customer-related proactiveness in the following sections. Holt *et al.* (1984: 29–30) advocate this kind of discussion and urge researchers to study proper needs-assessment methods for the different stages of the innovation development process. In particular, when proactiveness is considered part of *the process*, it is necessary to tie it to a certain objective, i.e. to define "toward what" the firm is proactive. This objective seems to vary at different stages of the innovation development process.

4.4 The role of customers at the idea-generation stage

4.4.1 Features of the idea-generation stage

It has been contended that the earliest phases of new-product development play a central role in the success of the product. Decisions made then are not easily changed in the later phases, because then the cost of and time required for any corrective action and engineering change increase considerably (Bacon *et al.* 1994: 40–45; Gupta and Wilemon 1990; Khurana and Rosenthal 1998). Nevertheless, it has also been argued that the early phase is not so crucial after all, since early decisions are often made without intensive anticipation, and binding choices are delayed to the point when the uncertainty has decreased (Eisenhardt and Tabrizi 1995). Therefore, developing a capability to create changes later, as a response to the more reliable new information, may lead to success in the development of new products (Ward *et al.* 1995) or innovations (Eisenhardt and Tabrizi 1995). However, the ability to incorporate later changes may vary by industry. Wyner (1998/1999: 49) argued that information-intensive industries (e.g. financial services and telecommunications) are more flexible than capital-intensive industries (e.g. the automotive industry).

The idea-generation stage begins with a search for or the picking up of ideas for a new product and ends with the decision to go further with the product development. It can be further divided into three interrelated steps: ideation, idea evaluation and the decision to approve funds for development (cf. Gobeli and Brown 1993: 41; Majaro 1992: 18; Trott 2002: 212).

Ideation means picking up or creating a large quantity of ideas, without paying attention to their quality (cf. Majaro 1992: 18). Stefik and Stefik (2006: 119) call this an "Aha!" moment. According to Gobeli and Brown (1993: 40), the ability to create new ideas is often restricted by a lack of resources, i.e. time, money and people. These resources tend to be directed toward current activities and not to the pursuit of new ideas. Furthermore, especially in human resources, quality rather than quantity shows in idea creation; the creative capability of individuals may differ (cf. Majaro 1992: 68–74).

Although the continuous, conscious, and purposeful search for ideas is imperative (Drucker 1985: 68), radical innovations often emerge at the periphery of the firm's field of vision, not at the centre, and thus may arise unexpectedly. However, this does not mean that they are arbitrary: they are often merely hidden in the plenitude of work done in the pursuit of some other invention (Hillis 2002: 152).

Ideas can be *evaluated* based on commercial, financial and technical criteria (Cooper, R. 2000: 114; Majaro 1992: 20). According to Jolly (1997: 6), this analysis tends to be highly subjective: some put more emphasis on technical merits, while others pay more attention to market potential. The evaluation of commercial feasibility seems to be especially hard in the case of radical innovations. The use of traditional marketing-assessment techniques is difficult, since customers have no experience with similar products. Furthermore, estimating future success in the market is troublesome, since the market may not even exist at the time (Deszca *et al.* 1999: 617). Nevertheless, it is important to be able to identify alternative market segments for multiple applications even at this stage (LaPlaca and Punj 1989: 99), otherwise the pressures on short-term financial performance may lead to the rejection of most radical ideas (cf. Deszca *et al.* 1999: 617). Evaluating technological feasibility is also problematic, since the technologies may also be at a very early stage of development at this point (Deszca *et al.* 1999: 618; Jolly 1997: 7).

The decision to proceed to product development implies the commitment of resources, both human and financial (Cooper R. 2000: 120–121). Implementation is of paramount importance, since "ideas are useless unless used" (Levitt 1963: 79). According to Levitt (1963: 81), although creativity is needed for creating ideas, implementing them requires discipline. It has also been emphasized in earlier literature that, in general, innovations arise from discipline more than from imagination (Drucker 1985; Pearson 2002).

Thus, even at the idea-generation stage, it is important to detect ideas that might eventually be turned into business: such ideas are called opportunities (Allen 2003: 27). The following section discusses how these opportunities may be recognized or created.

4.4.2 Customer-related proactiveness at the idea-generation stage

Customer-related proactiveness and reactiveness at the idea-generation stage seem to be connected to market opportunity, i.e. whether the opportunities are reacted to (recognized in the marketplace) or anticipated or created (cf. Allen 2003: 27).

A situation in which opportunities are created by the firm reflects high customer-related proactiveness, which often seems to characterize the generation of ideas for radical innovations. In other words, according to the study conducted by Veryzer (1998b: 142), ideas for radical innovations tend to be generated internally and are often driven by the desire to apply a particular technology (see also Holt *et al.* 1984: 7). This indicates that basic research often plays a significant role in the emergence of radical innovations (see also Rogers 1983: 138–139; Veryzer 1998a: 313). Basic research is directed to the advancement of scientific knowledge without any specific objective of applying this knowledge to practical problems (Rogers 1983: 138; Trott 2002: 302). It involves the exploration of various technologies in R&D laboratories and the generation of "solutions searching for problems." This drifting may go on for years until the necessary convergent technological advances are made (or abandoned). Understanding how all the pieces fit together into a particular application of the technology requires both good technological competence and a general feeling for the market (Veryzer 1998a: 313–315).

The benefit of scientific knowledge gained as a result of basic research can often be felt later on in the applied research (Rogers 1983: 138). Applied research uses existing scientific principles to solve a particular problem (Trott 2002: 302). Thus, when the ideas for radical innovations stem from inside the firm, from applied research that originates in basic research, it could be said that *the firm itself creates opportunities in the market, i.e. behaves very proactively toward customers* (see Figure 4.3).

The results of previous studies also suggest that opportunities in the market are often *anticipated* at the idea-generation stage, and this is also indicative of proactiveness. A scientist may launch a research project to find a solution for an anticipated problem, and thus react to future needs (Rogers 1983, 135; cf. Figure 3.2). Furthermore, applied research may stem from the perception of the customers' existing or latent needs. However, in the idea generation of radical innovations, even customers' existing needs tend to be rather vaguely expressed. Since they rarely understand what is possible, it is often difficult for them to articulate their needs (Kaplan 1999: 21). Consequently, previous studies (e.g. Kaplan 1999; Lawton and Parasuraman 1980; Millier 1999: 11, 15) argue that radical innovations are not usually born by *reacting* to specific desires expressed by customers.[6]

Nevertheless, von Hippel *et al.* (2000: 22) emphasize the role of customers. Suitable methods for anticipating market opportunities for radical innovations seem to include visioning, lead-user analysis, customer immersion[7] and empathic design[8] (Deszca *et al.* 1999). Tools used in visioning tend to be less

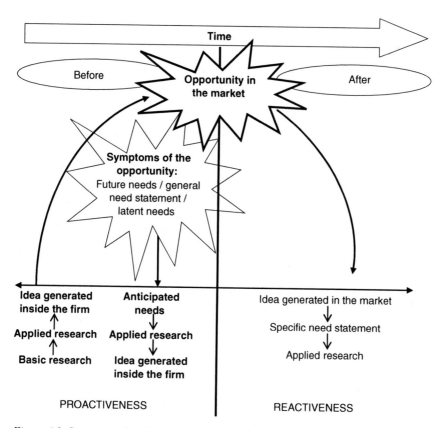

Figure 4.3 Customer-related proactiveness and reactiveness at the idea-generation stage.

structured than those used in searching for ideas for incremental innovations (O'Connor and Veryzer 2001: 244). For instance, scenario mapping, the Delphi method and back casting also seem to be of use in the case of radical innovations (Deszca *et al.* 1999: 621–622; O'Connor and Veryzer 2001: 232–233).

Von Hippel *et al.* (2000: 22) argue that, by identifying lead users and learning from them, firms may systematize the creation of radical innovations and better ascertain the commercial success of the product. Lead users may be defined as users whose current strong needs will become general needs in a marketplace in the future (von Hippel 1989: 24). Thus, they seem to be an ideal source for innovative ideas. This notion is not usually adopted as such, however, but rather combined with ideas from other lead users and in-house developers (cf. Veryzer 1998b: 142; von Hippel *et al.* 2000: 23). In general, the utilization of lead users seems to be more applicable to the development of incremental rather than radical innovations (Olson and Bakke 2001: 392–393). Furthermore, it is rather demanding to identify lead users whose current needs match the future needs of mainstream customers (Mohr 2001: 150–151; Moore 1999b: 16; Urban and von

Hippel 1988: 577–581). It would therefore be important to hold direct discussions with mainstream customers and to gain deeper knowledge of their behavior and potential future needs, provided that they are able to assess them (Deszca *et al.* 1999: 624).

Consequently, it seems that, particularly with radical innovations, *customer-related proactiveness tends to be either very high (influencing) or high (anticipating) at the idea-generation stage.* This corresponds to the earlier discussion suggesting that they create their own market. Ideas are usually generated inside the firm, and they originate either from anticipated needs or from basic research.

Provided that the innovation really is radical, these opportunities are likely to be global. In other words, if the innovation creates substantial benefits for the customers, and if it is so new that there are no competitors, one could conclude that the opportunities are not limited to one country, or even to one region (cf. Christensen 1997; Nayak and Ketteringham 1986). They may even be found in the most attractive niches around the world, where the first mover might be able to limit the amount of space available for imitative later entrants (cf. Lieberman and Montgomery 1988: 44–45). Thus, the scope of the firm's customer-related proactiveness tends to be *geocentric* at this stage.

4.5 Developing the innovation to meet customers' needs

4.5.1 Features of the development stage

The development stage could be defined as the process of putting a new idea into a physical and psychological entity that is expected to meet the needs of an audience of potential adopters (Rogers 1983: 139–140; Urban and Hauser 1993: 39). In terms of product innovation, this means deriving the details and specifications for the proposed product and making sure that it functions properly with other parts of the system into which it will be integrated (Cumming 1998: 23). With service innovations, the development stage comprises the design and creation of activities and processes that lead to the new service offering (cf. Avlonitis *et al.* 2001: 325–328). The development of R&D and technology-intensive services resembles the development of technology-intensive products and vice versa, the role of services is often accentuated in the development of technology-intensive products (cf. Miles 1994: 252). Especially in radical innovations, consideration of the product's support requirements is often closely integrated into the development stage. Product support includes the assistance that is offered to customers to help them gain maximum value from the innovation and is therefore related to customer satisfaction (Goffin 1998: 42–43).

Problems at the interface of the marketing and engineering functions may hinder the firm from completing the new-product design for marketing in the development stage (Gobeli and Brown 1993: 42). Many studies have shown that that the integration of marketing into the development process enhances new product success (e.g. Ettlie 1997; Hart *et al.* 1999; Hise *et al.* 1990). Souder *et al.* (1998: 522) argue that in radical innovations in particular, a high degree of

R&D/marketing integration seems to be required in order to overcome the difficulties caused by high market and technological uncertainties. Nevertheless, as Veryzer's (1998b) study indicates, the development of radical innovations is driven and controlled by R&D managers, and marketing personnel are involved later than in the case of incremental innovations. This is a rather natural consequence of the fact that radical innovations are usually derived from the R&D environment, and technology often drives the development until a viable product application can be determined. Be that as it may, the earlier integration of marketing into the development process could be beneficial (Veryzer 1998b: 148).

Close relationships with customers may also improve the innovation development process. Veryzer's (1998b: 149) study also shows that partnerships with key customers occasionally play an important role in developing radical innovations. If customers are involved in the development, the offer and need tend to converge earlier than is the case in closed-product development. Besides, customers participating in the development tend to be much more inclined than others to adopt the innovation later on when it is launched (Millier 1999: 82–86). However, especially in the case of radical innovations, the use of customer involvement may be impeded by its drawbacks, i.e. information leaks and ownership problems (cf. Millier 1999: 96–97). Building and maintaining the customers' trust is therefore often seen as a prerequisite for involving them and disclosing knowledge to them (cf. Hurmelinna *et al.* 2002).

Since customers have no experience with similar products, it is difficult for them to define the product specifications (Deszca *et al.* 1999: 618). However, a recent study by Lettl *et al.* (2006) suggested that when customers are professional, know the technology and work in the field, they may be able to make active development contributions even for radical innovations. Usually, however, in order to get customer feedback, it is necessary first to educate them about the products and the usage contexts. Consequently, the acquisition of customer feedback may become costly and time-consuming (Deszca *et al.* 1999: 618).

Particularly in radical innovations, the role of customer feedback seems to vary depending on whether the prototype is ready or not (Trott 2002: 269; Veryzer 1998a). Given the evident importance of prototype, the following sections examine customer-related proactiveness separately in two stages: pre- and post-prototype.

4.5.2 Customer-related proactiveness before the prototype construction

The earliest phases of innovation development are the most precarious: an arena of vast tacit knowledge, conflicting organizational pressures and considerable uncertainty. It would therefore be beneficial to know the market needs even at this stage (see also Bacon *et al.* 1994; Khurana and Rosenthal 1998). According to Cumming (1998: 23), the failure to understand what the market wants will almost certainly damage the prospects of the product, although there is no guarantee that even a good understanding will lead to success. Several studies on new product development have thus highlighted the important role of market information (e.g. Cooper 1979: 100–101; Hart *et al.* 1999: 20; Ottum and Moore 1997; Urban and Hauser 1993: 141).

Market information is required in order to develop user-relevant product advantages (Hart *et al.* 1999: 20). Since customers will eventually buy "a bundle of benefits encompassing the physical product and its extended service offerings" rather than a set of product features, it is critical to ensure that the product will really create value for its target segments (Wind and Mahajan 1997: 7). If the firm fails to take the end users into consideration early in the innovation process, it runs the risk of producing mere technical devices, i.e. devices that have unnecessarily high performance levels but no real value to the customers (Millier 1999: 43–46). Thus, customer feedback improves the chances of making the right decisions along the developmental path (Veryzer 1998b: 136). Furthermore, market information can also help to accelerate the development process, thus closing the time gap between idea generation and launch[9] (Mohr 2001: 116, 118).

Market information clearly reduces uncertainty and risk in the development of new products (Hart *et al.* 1999: 22). It is particularly important to a technology-based firm, as "this information shapes science into a commercial product or service" (Leonard-Barton 1995: 177; cf. Sultan and Barczak 1999: 25). Even though the need to incorporate customer information into the product development process is known, there is a common tendency to overlook it. It is not easy for customers to envisage how the innovation will meet their needs: after all, they might not even know what needs they have (Mohr 2001: 116), or they might not be able to articulate them clearly (von Hippel 1986: 802).

In the case of radical innovations, which tend to be rather loosely defined,[10] knowing market needs is often not even possible in the early stages (cf. Bacon *et al.* 1994; Khurana and Rosenthal 1998*). It is hard to obtain meaningful customer input before the prototype is ready, since it is difficult for customers to imagine the radical innovation or the implications it carries for their businesses or lives.* Since customers often have no experience of the technologies underlying these products, they have little or no frame of reference for understanding them. Even though it may be possible to educate them about the coming radical innovation as part of the market research, this may skew research findings (Veryzer 1998b: 143).

At this stage, a firm may try to collect customer information via information acceleration (cf. Deszca *et al.* 1999: 623). This is a method that, according to Wyner (1999: 36), "tries to bring the future into the present" and involves simulating alternative environments and measuring customer responses to them. Thus, customer attitudes toward electronic presentations of product variations can be analyzed (Deszca *et al.* 1999: 623). However, even though information acceleration often provides valuable data for product developers, it is also a rather expensive and demanding technique, which can never fully simulate the real decision environment of the customer (Urban *et al.* 1996).

Due to the difficulties in conveying to potential customers a real sense of radicalness, various researchers (e.g. Christensen 1997: 147–162; Holt *et al.* 1984: 30; Leonard-Barton 1995: 204; Veryzer 1998a: 315) have questioned the value of collecting meaningful customer data at this point, since it may be a waste of scarce resources. Furthermore, it has been argued that too early anticipation of customer needs may actually discourage radical innovations (O'Connor 1998: 153). In any case, there is often a need to develop the product application before the product concept or target market is determined. Instead of concentrating on customer benefits and commercial opportunities, attention is rather paid to the technical differential advantage that the innovation will offer over existing products and technologies (Veryzer 1998a: 315–319).

At the early stage, general information about the market can be obtained from trend-watcher organizations and research institutes (Veldhuizen *et al.* 2001: 938), whereas *customer research is affected by secrecy concerns and the proprietary nature of radical innovations* (Veryzer 1998b: 142–143). The amount of information that can be provided to customers during the development process is limited. In any case, the extended length of the development cycles for some of these products further accentuates the need for secrecy in order to avoid alerting potential competitors to the existence of the development projects and their direction. Thus, the usability of customer research is restricted at the beginning of the development stage (Veryzer 1998b: 142–143).

According to Veryzer (1998b), relatively little formal customer research has been conducted at the beginning of the development stage. Understanding of the likely users and their needs has mainly been gained through observation and the immersion of development-team members in the user environment. The amount of research is rather limited, and there seems to be no consistent focus on customer issues. Furthermore, accurate perceptions of radical innovations have apparently been hard to obtain because they often involve complex connections with other aspects of customers' lives or businesses and change both their thinking and their behavior (Veryzer 1998b: 142–143). O'Connor's (1998) study gave even more drastic results. At this stage, there seemed to be "no foray into the market, no customer contact, no concept test with lead users. Rather, there is a period of technological forecasting coupled with imagination or visioning" (O'Connor 1998: 158).

When product specifications are further developed, some information is also collected on use requirements. Customer-research methods applied at this stage are rather informal and include observation or internal product tests (Veryzer 1998a: 315). A cursory market analysis is often carried out before the prototype construction begins. This analysis may meet formal project-review requirements and may also be motivated by the need to show that the project is worth continuing and that the product might be commercially viable. It is necessary to carry out this kind of evaluation before the project has absorbed considerable amounts of research and resources. The product concept plays an important role in the evaluation: it captures the original idea and also acts as an integrative point of engineering, feasibility and market analysis. The market analysis identifies

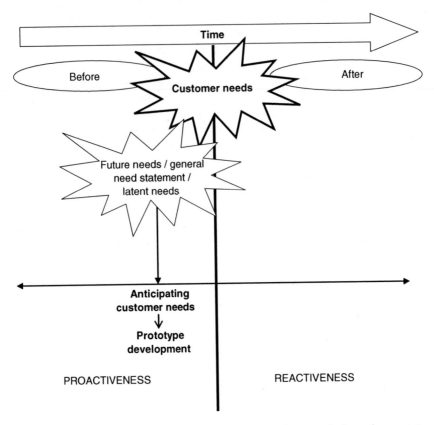

Figure 4.4 Customer-related proactiveness and reactiveness before the prototype
     development.

target markets and produces sales projections. Naturally, there is high uncertainty associated with it at this stage (LaPlaca and Punj 1989: 103–104; Veryzer
1998a: 315), which suggests that customer research before the prototype construction may be rather problematic and vague in the case of radical innovations
(Veryzer 1998a: 319; Veryzer 1998b: 143).

   In sum, customer-related proactiveness and reactiveness at the development
stage seem to be connected with customer needs, i.e. whether these needs are
reacted to, anticipated or influenced (see Figure 4.4). *Before the prototype is
developed, the needs of customers are* not actively influenced, but are rather
anticipated, to some extent (O'Connor 1998; Veryzer 1998a).

   However, customer-related proactiveness seems to change considerably
during the prototype-construction phase. This change is described in the following section.

4.5.3 Customer-related proactiveness after the prototype construction

Once the product application has been formulated and the prototypes developed, the development process tends to become more customer-oriented and more customer input is used. According to Veryzer (1998a: 316), this is the point at which the process begins to follow the lines of more conventional new-product development. The production of a prototype enables the firm to give its customers a true sense of the innovation (its features, benefits and capabilities) for the first time and to assess their reactions. This helps it to further determine the specifications and to assess market attractiveness and financial expectations (Veryzer 1998b: 142). It is therefore no wonder that firms developing radical innovations tend to develop prototypes at an earlier stage than those developing incremental innovations (O'Connor 1998: 160; Veryzer 1998a).

It is clear that understanding customer needs and specifications is crucial at this phase, given the high uncertainty involved in radical innovations and the unfamiliarity among customers. Customer-research activities, first rather informal and qualitative, and later more formal and quantitative, are thus initiated (LaPlaca and Punj 1989: 105; Veryzer 1998a: 316, 319; 1998b: 142). These activities are designed to reduce the uncertainties inherent in the new product development process (Hart *et al.* 1999: 24).

A variety of different marketing-research techniques are used, depending on the nature of the innovation and the type of customer data that is considered suitable for evaluating it (Veryzer 1998b: 142). Customer tests may be conducted with lead users or key customers (O'Connor 1998: 160; Veryzer 1998a: 316). Field-testing of prototypes helps to clarify what adjustments are needed to enhance marketability, and test-marketing may assist in refining estimates of customer response and the marketing effort required (LaPlaca and Punj 1989: 104–105). If the product cannot be produced for test markets, the firm has to rely on consumer-based laboratory simulations of the buying process (Urban and Hauser 1993: 39).

Thus, the prototype formulation has an influence on the degree of proactiveness. Before its development, the needs of the end market are anticipated rather than actively influenced and it is later that prospective customers are usually better capable of expressing their needs in terms of the innovation and its features (see Figure 4.5). Thus, *customer-related proactiveness tends to decrease as the development proceeds further toward the launch stage* (O'Connor 1998; Veryzer 1998a).

Studies on innovation development suggest that, during the whole development stage, the needs of the end user are anticipated (pre-prototype) and reacted to (post-prototype) mainly in the context of the home market. The international aspect seems to be difficult to accommodate, since the communication includes the exchange of a considerable amount of tacit knowledge, the cost of which increases rapidly with the spatial dispersion of people and research units (Lindqvist *et al.* 2000: 117–120). In any case, secrecy concerns and expense often prevent the conducting of market research abroad (cf. Devinney 1995: 74).

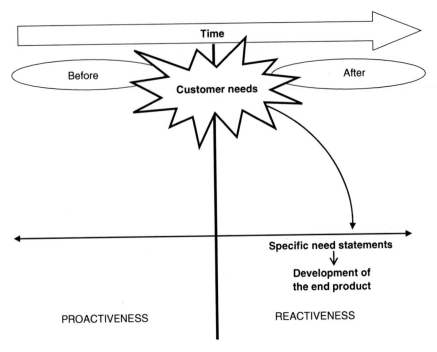

Figure 4.5 Customer-related proactiveness and reactiveness after the prototype development.

Thus, although the global potential of the innovation is prevalent at the idea-generation stage (see Section 4.4.2), this potential may even be vitiated because of local centralized development (cf. Hedlund and Ridderstråle 1995: 165). Hedlund and Ridderstråle (1995: 168) called international development projects "glocal," meaning that they are global in the beginning and local later on. Customer-related proactiveness at the development stage thus seems to be *ethnocentrically* oriented.

4.6 Assuring customer acceptance for the launch

4.6.1 Features of the launch stage

It may be hard to distinguish when the innovation development process moves from the development to the launch stage. These phases are distinguished in terms of their emphasis in this study: whereas the emphasis at the development stage is on matching the innovation with customer needs, and the focus is largely on its features, it turns at the launch stage toward interesting prospective customers in the innovation and in buying it.

Thus, the launch stage starts before the actual launch takes place, i.e. when the firm begins to plan it (cf. Hultink *et al.* 1997: 245), and it ends when the innovation has been successfully launched onto the market. Successful launch means that the innovation has become established in the market (Urban and Hauser 1993: 40)[11]. In the case of radical innovations, establishment implies adoption and diffusion (Guiltinan 1999: 511–512), meaning that it does not suffice for only innovative adopters[12] to acquire the innovation and that diffusion has to spread to the mainstream market (Moore 1999a: 28–52).

A radical innovation is introduced onto the market at the launch stage. According to Urban and Hauser (1993: 39), launching is "the difficult task of 'making the product happen' in the market." Guiltinan (1999: 509) states that the launch stage aims at maximizing the chances of profitably gaining acceptance in the target market. Furthermore, Song and Montoya-Weiss (1998) found that activities related to coordinating, implementing and monitoring the launch tend to have very strong effects on innovation success. Other researchers (e.g. Beard and Easingwood 1996: 88; Easingwood and Koustelos 2000: 33–34) have also emphasized the critical role of the launch stage. Investment in the innovation may be considerable by this time, and the risk of rejection is still quite high. This makes the manner in which the launch is handled important (Easingwood and Koustelos 2000: 33–34). It is therefore worth analyzing how customer-related proactiveness may contribute to customer acceptance.

4.6.2 *Customer-related proactiveness at the launch stage*

Given the above-mentioned significance of the launch stage, it seems rather natural to assume that *anticipation* should play a role in ensuring that the innovation will be accepted. Thus, customer reactions should be anticipated through market research before the launch (Di Benedetto 1999: 539). Customer research conducted at this point aims at further refining the innovation features and at clarifying marketing-related issues (e.g. target market, pricing). Research methods tend to be more formal and samples larger than before (Veryzer 1998b: 142).

The situation may be quite different for radical innovations, however. According to a study conducted by Hultink *et al.* (1997), the degree of product innovativeness influences the decisions made in the launch process. Consequently, it has also been argued that it may sometimes be better with very radical innovations to be faster onto the market and to strive for only minimal acceptability rather than to pursue a complicated combination of not-so-relevant product attributes (cf. Lieberman and Montgomery 1988; Song and Montoya-Weiss 1998: 131–132). After all, detailed market studies on radical innovations are extremely costly and still do not guarantee success (Song and Montoya-Weiss 1998: 132). According to Beard and Easingwood (1996: 87–103), firms tend to make modifications in their radical innovations as a *reaction* to customer feedback during and after the launch. This seems rather natural, since the information becomes more valid and reliable as the innovation gets closer to the market (see Crawford 1997: 381–400).

The process of refining the innovation after the launch, as the firm learns from customer reactions, has been called "the probe and learn process" by Lynn *et al.* (1996). According to their study, probing and learning seem to be characteristic of firms launching radical innovations. Instead of analyzing the market in detail and carefully selecting the best alternative, these firms introduce an early version of the innovation, learn from the feedback and make appropriate modifications (Lynn *et al.* 1996). Mullins and Sutherland (1998) studied the process of new product development in rapidly changing markets (e.g. telecommunications and computer software) and found out that, in order to reduce uncertainty and mitigate risk, these firms tend to rely on probing and learning. The products were introduced quietly, without massive launch campaigns, and the product offerings were later updated to better meet customer requirements (Mullins and Sutherland 1998: 232–233).

Continuous monitoring and reacting to customer requirements is necessary, because market preferences may change in the period between the original design specification and the launch. Innovation developers are thus aiming at a moving target, and customer research gives only a temporary overview of the continuously changing market (Millier 1999: 100).

Complementing the above discussion on the importance of reaction at the launch stage, previous research has also emphasized *influencing* customers and preparing them for the innovation. Past studies have identified three basic ways in which firms launching radical innovations are inclined to influence their customers. First, *awareness building* is stressed particularly strongly (e.g. Easingwood and Koustelos 2000: 27–28; Guiltinan 1999), although getting customers to understand how an innovation works or giving them information on its as-yet-unobservable benefits does not make it less complex or its benefits more apparent. It is rather the pre-launch information focusing on relative advantage and compatibility that seems to have the biggest direct influence on trial and adoption. Positioning the benefits against those of other products is likely to improve understanding in this sense (Guiltinan 1999: 514). Trade shows are also frequently utilized to communicate benefits of new innovations and to create publicity (Uslay *et al.* 2004: 20).

Second, the more radical the technological capability inherent in the innovation is, the greater is the need for *customer education*: with radical innovations, market education may be even more strongly directed toward communicating a vision of the future than toward the nature of the technology itself (Beard and Easingwood 1996: 94–100; cf. Easingwood and Koustelos 2000: 28; Song and Montoya-Weiss 1998: 132).

Third, *giving customers the opportunity to try the innovation* before buying it seems to be especially important in radical innovations. If this is not possible, word-of-mouth communication between early adopters and prospective customers seems to partly compensate (Hoeffler 2003: 415–418).

It is important to consider the real needs of the end customers when the firm is preparing the market. Rackham (1998) argues that one of the mistakes usually made during the launching of radical innovations is that when the products are introduced to the sales force (which influences the way in which products are sold), the focus is usually on all of their innovative elements, which is likely to turn the salesperson's attention away from the customers' needs. This makes the selling of innovations product rather than customer-centered, which may cause customer interest to fade. After all, customers pay attention to the capacity of the innovation to meet their needs, and consider its features only on this basis (Rackham 1998: 201–206).

In sum, at the launch stage the key issue seems to be that the customers accept the new innovation (see Figure 4.6). This compatibility between innovation and customer seems to be achieved by influencing the customers (i.e. through awareness building, education and trial marketing) and by making customer-required modifications after the launch. Thus, *both customer-related proactiveness and reactiveness tend to be high at this stage*.

A large market is needed to cover the costs and risks associated with the development of a radical innovation (Oakley 1996: 79; cf. Ohmae 1989: 153–155). Furthermore, as knowledge of a new innovation spreads around the globe shortly after its discovery (Dekimpe *et al.* 2000: 49), it is important to be able to launch the innovation simultaneously in various countries. In order to be

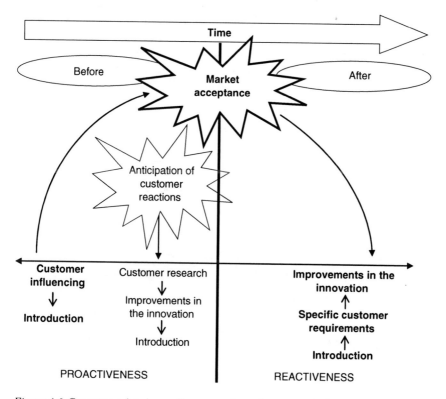

Figure 4.6 Customer-related proactiveness and reactiveness at the launch stage.

able to use the limited "window of opportunity" before competitors or changing customer preferences close it, it is thus often necessary to introduce the innovation rapidly onto global markets (Hurmelinna *et al.* 2002). This means that a geocentric orientation may be the most appropriate option at this stage. This is further supported by the results of studies (Kleinschmidt and Cooper 1988; Oakley 1996) indicating that more successful launches are strongly associated with faster introduction onto overseas markets and a higher proportion of international sales. Naturally, however, a fast overseas launch alone does not guarantee success if it is not allied to a well-planned marketing effort (e.g. carefully timed customer preparation) and effective distribution (cf. Devinney 1995: 72; Oakley 1996: 87–89).

Although a geocentric orientation seems to be beneficial, the need for market preparation often leads to *polycentric orientation* in practice. At the launch stage, this could take place through both sequential entry and adaptation of the launch concept (see Figure 4.7).

According to Mascarenhas (1992), sequential entry into the most important markets tends to take place with novel products, whereas at the later stages of the product life cycle, market entry tends to become more simultaneous and oriented toward smaller catchment areas. Furthermore, Hu and Griffith (1997: 119–122) argue that radical innovations often require extensive marketing adaptation in each country at the beginning of their launch, and it is only when they start to gain acceptance that the regiocentric or geocentric approach is appropriate. Consequently, the sequential and adapted approach tends to be more common in innovations that are difficult to adopt in local markets, e.g. those that are technologically too complex to be handled by the local sales force (Chryssochoidis and Wong 1996: 188).

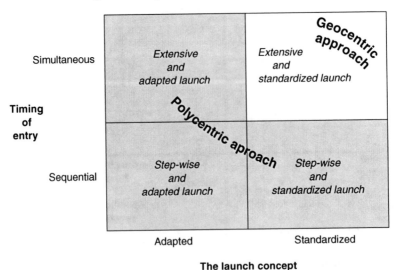

Figure 4.7 The polycentric versus the geocentric approach in the launch.

The sequential and adapted approach is also required in innovations that have to undergo onerous approval procedures (Chryssochoidis and Wong 1996: 185). This seems to be the case in many radical innovations, when the launch often depends on external country-related factors such as obtaining patent rights and selling permission (cf. Devinney 1995: 72). Consequently, their diffusion may often be more decentralized than centralized. In the latter case, the firm makes the decisions on where to sell the innovation, whereas in the former, the local units (e.g. governments) determine whether or not to allow the diffusion: in this case firms may naturally try to influence these decisions (Rogers 1983: 334–342; Dekimpe *et al.* 2000: 58).

Moreover, the sequential approach is further favored when there are not enough resources to launch the innovation in many countries simultaneously. Although the restrictions are often financial, they may also be related to knowledge, such as in a situation in which the sales force is not familiar with the local target groups (Acs *et al.* 2001; Chryssochoidis and Wong 1996: 189). Overall, target groups seem to be influential in both local and international markets. The following section deals with targeting.

4.6.3 Adjusting customer-related proactiveness to different target groups

As stated above, proactive behavior tends to be expensive. The costs of customer-related proactiveness are more justifiable if it could be targeted at certain groups of prospective customers. Moreover, the results of previous studies indicate that targeting increases the diffusion rate provided that the firm finds the right people: targeting the wrong people may inhibit diffusion completely (Dunphy and Herbig 1995: 196–200; Gatignon and Robertson 1985).

Innovative adopters[13] seem to be a natural target group for most radical innovations at the early launch stage and comprise two different groups of people: innovators and early adopters (Easingwood and Koustelos 2000: 29; Rogers 1983: 247). These groups are united by their desire to be the first, either to explore (innovators) or to exploit (early adopters) the new innovation. By being the first to exploit they aspire to achieve considerable advantage over the old regime. Given this desire, the competition at the launch of a radical innovation is not among alternative products but rather among alternative radical solutions (Moore 1999b: 15–16, 176).

Innovators appreciate innovativeness for its own sake and are genuinely interested in learning about new products and services. They enjoy mastering the intricacies of the new technology and wish to try it just in order to see whether it works. Thus, they are adventurous and also accept setbacks. They are influencers and play a gate-keeping role in the innovation–diffusion process, which makes them a key target group (Moore 1999b: 15; Rogers 1983: 248).

Early adopters are visionaries who are looking for breakthrough improvements in their current productivity and customer service. They want to be the first to exploit the innovation and thus achieve new competitive advantage. They are not very price-sensitive, which makes them an easy group to sell to. Moreover, they are often very willing to serve as very visible references. After all, these people are respected among their colleagues because of the successful use of new ideas, and in order to guarantee this esteem in the future, they are willing to make decisions on radical innovation adoption earlier than their peers. However, they represent a challenge to the innovative firm in that they are also very demanding in terms of both innovation properties and time frame (Moore 1999b: 15–16; Rogers 1983: 248–249).

It is worth emphasizing the fact that innovators and early adopters require partly different approaches to customer-related proactiveness. Customer preparation is used to make them aware of the innovation and skillful in using it properly. Building on awareness is rather easy among this group, since they actively seek knowledge of new radical innovations. As Moore (1999a: 36) writes: "Typically you don't find them, they find you." Both innovators and early adopters are often eager to participate in conferences, seminars and trade fairs but require different programs. The former are eager to know more about the technical capabilities, whereas the latter want reliable evidence of the promised overall benefits. Consequently, these same differences should be reflected in the education of early innovators about the innovation, which will be effective as long as it emphasizes the technical details for the innovators and the potential for the early adopters (Beard and Easingwood 1996; Easingwood and Koustelos 2000: 29; Moore 1999a: 28–36).

Pre-launch market research targeted on innovative adopters is likely to be rather stimulating since they eagerly participate in the product development and are particularly pleased to play a part in the innovation process. On the other hand, market research may also be rather demanding because early innovators in particular often insist on various modifications, which quickly begin to tax R&D resources (Moore 1999a: 34–35).

The post-launch benefits and drawbacks in terms of reacting to the requirements of this group are rather similar to when customer research is conducted before the launch. However, early innovators usually do not represent typical customers, and it might therefore be more beneficial to target the mainstream market, especially the early majority, afterwards (Mohr 2001: 150–151; Moore 1999a: 39–40; Moore 1999b: 16). Increasing evidence suggests that the needs of the early majority are not the same as those of innovative adopters (Mohr 2001: 150–151; Moore 1999a: 55–57; Wind and Mahajan 1997: 7), and it is thus important not to be blind-sided by the enthusiasm and ease with which the latter might use the product (Mohr 2001: 150).

The early majority[14] comprises customers who approach innovations with cautiousness (Rogers 1983: 249) and who look for proven and reliable applications rather than technological breakthroughs (Mohr 2001: 151; Moore 1999b: 16). They can be influenced through the social system, i.e. by utilizing references and trusted peers (Moore 1999a: 42). It is worth emphasizing that these customers do not trust innovative adopters but rather rely on people coming from their own group, i.e. from the early majority (Moore 1999a: 42).

In order to convince the early majority of the viability of the innovation, it is often necessary to react to their requests and develop modifications accordingly (Moore 1999a: 43–44). However, when the demand starts to grow very quickly, there may be no room for reaction: "[The customers] do not need – or want – to be courted – the problem is not to create demand: they need – and want – to be supplied!" (Moore 1999a: 79). Hence, it seems from previous studies that customer-related proactiveness can be targeted toward different groups of adopters at the launch stage.

4.7 A framework of customer-related proactiveness during the development process

A framework for analyzing the change in customer-related proactiveness during the development of radical innovations is presented in Figure 4.8. It is based on Pettigrew's (1988) suggestion that longitudinal field research on organizational processes should include not only the change process itself but also the content and contexts of the change and their interconnections over time. Thus, the inner and outer contexts both appear in the framework, even though the main emphasis is on change in customer-related proactiveness.

The degree of proactiveness toward customers is described in the framework in terms of its position on the reactiveness–proactiveness continuum at the idea-generation, development and launch stages. Since both concepts always seem to necessitate an objective, the key issue toward which the firm either pro-acts or reacts, it was rather natural to specify this key issue for each stage and thus to utilize the continuum to describe the degree of proactiveness. In fact, without these key issues, analysis during the process would have been impossible, since the time dimension is inherent both in the innovation development and in the very concept of proactiveness.

Furthermore, the international scope of proactiveness is circled in the framework: ethnocentric scope (E) in a small circle, polycentric scope (P) in a larger circle, and geocentric scope (G) in the largest circle.

According to previous studies, customer-related proactiveness is usually high or very high at the idea-generation stage, particularly in radical innovations. In other words, innovations create their own markets by either directly creating or anticipating opportunities. The scope of the firm's customer-related proactiveness seems to be geocentric at this stage, since the opportunities for radical innovation are likely to be global.

At the development stage, customer-related proactiveness and reactiveness seem to be connected to customer needs, which are often anticipated before the prototype is ready. However, proactiveness seems to change during the prototype-construction process and seems to decrease as it gets closer toward the launch stage. Reacting is possible, since having a prototype often enables customers to better express their needs regarding the innovation and its features. Furthermore, customer-related proactiveness at the development stage seems to be *ethnocentrically* oriented, and end-market needs are anticipated

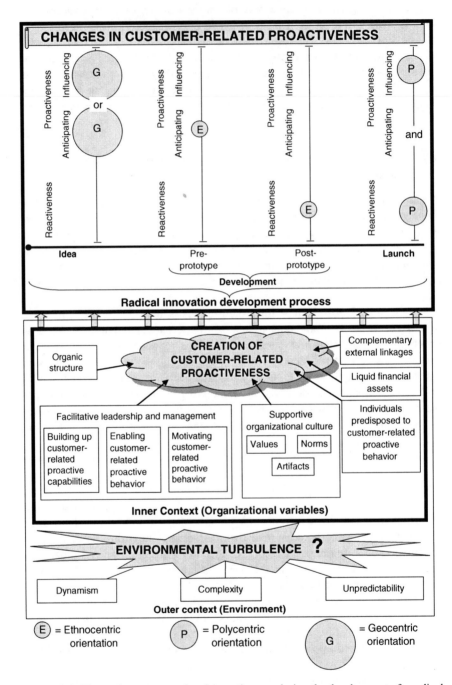

Figure 4.8 Change in customer-related proactiveness during the development of a radical innovation.

(pre-prototype) and reacted to (post-prototype) mainly based on the home market.

At the launch stage, compatibility between the innovation and the market seems to be achieved to a large extent by preparing the market and by making customer-required modifications after the launch. Consequently, both market proactiveness and reactiveness tend to be high at this stage. In any case, the need for market preparation often seems to lead to a polycentric orientation (i.e. sequential entry and adaptation of the launch concept).

Thus, it seems that both the nature and scope of customer-related proactiveness change during the process of radical innovation development. Different stages require or allow different kinds of proactiveness, but it does not occur naturally or automatically in an organization: on the contrary, the creation of customer-related proactiveness seems to require the interplay of various organizational variables. According to the literature, it seems to be fostered by an organic organizational structure, facilitative leadership and management, the organizational culture, individuals predisposed to this kind of behavior, liquid financial assets and complementary external linkages.

It is not only organizational factors but also perceived environmental turbulence in the form of environmental dynamism, complexity and unpredictability that may influence proactive behavior. Previous studies have shown rather contradictory results in terms of whether the higher level of perceived turbulence increases or decreases it, however, which is why the role of environmental turbulence is questioned in the framework.

It is worth emphasizing that the framework presented in Figure 4.8 is based on the existing literature, which is rather deficient in terms of specific customer-related proactiveness. It is preliminary and descriptive rather than predictive. It functions here as a preliminary conceptualization of the nature and scope of customer-related proactiveness and the factors fostering it and will be further complemented and specified based on the case studies.

5 Customer-related proactiveness in the case innovations

This chapter first describes how the empirical study was conducted. The main emphasis is on how the case innovations were developed and on the degree and international scope of customer-related proactiveness in each development process. Unless otherwise stated, the descriptions are based on the information given in the interviews and on discussions and correspondence (Appendix 2), and complemented and verified by means of secondary and company documents (Appendix 3). Although the detailed analysis of each innovation process covers the period up to 2002, it also gives information on how the resulting innovation has fared since then.

5.1 Empirical research design

5.1.1 Pilot study: Domosedan

The study began with a *pilot case study*, which was used to evaluate the usability of the a priori framework and the operational measures used. The pilot study described the development of a radical pharmaceutical innovation, Domosedan®. Since the pilot study has been reported in the *Journal of Targeting, Measurement and Analysis for Marketing* (Sandberg 2002)[1], in the *European Journal of Innovation Management* (Sandberg and Hansén 2004)[2] and in the *Publications of the Turku School of Economics and Business Administration* (Sandberg 2004)[3], the brief description here is limited to its contribution to the focal study.

The most important consequence of the pilot study was the decision to limit the focus of the main study to customer-related proactiveness, instead of market proactiveness; hence, competitor-related proactiveness was not included. This decision was justified because both the degree and the international scope of market proactiveness seemed to differ considerably in customer-related and competition-related comparisons (Sandberg and Hansén 2004). Consequently, even though it would have been interesting to include both aspects in the focal study, it would have been too much to handle in one thesis. It was also acknowledged that including both aspects would have limited the number of cases to only a few. Therefore, it was decided to include more cases and to focus on the aspect that was often cited as the challenge for innovative firms, i.e. proactive behavior toward customers (cf. Mullins and Sutherland 1998).

The decision to concentrate on customers induced some modifications in the theoretical framework, which otherwise seemed to work relatively well. The results of the pilot study indicated that both the degree and the international scope of market proactiveness tend to change along the process of innovation development. Furthermore, the pilot study gave valuable experience about interviewing and relevant lines of questioning.

To conclude, the case study was valuable in guiding both the modification of the theoretical framework and the conducting of the empirical research. Once the pilot study was finished, the data collection for the actual case studies began.

5.1.2 Case selection

This study involved *multiple cases*, which helped in obtaining a more extensive understanding of the phenomena in different radical innovation development projects. It has been argued that multiple cases tend to allow the use of cross-case analysis for richer theory building (Perry 1998: 792), and to increase external validity and guard against observer bias (Leonard-Barton 1990: 250). The trade-off, however, is the consequent lack of depth in the individual case descriptions. Moreover, obtaining a holistic view of a specific phenomenon is often a time-consuming effort, and it is possible to carry out only a very limited number of case studies in one research project (Gummesson 1991: 76).

The cases for this study were selected according to *literal replication* logic, i.e. given the expectation that they would support the a priori framework (cf. Yin 1989: 53–58). Since previous studies on customer-related proactiveness were few and far between, there was not yet enough of a basis for theoretical replication. The selection of cases may be further constrained by research resources and accessibility (cf. Rowley 2002: 19). In this study, these constraints limited the choice to *Finnish* radical innovations. Possible cases were mapped through expert interviews and a brief executive survey.

First, the researcher conducted *two expert interviews*, aimed at finding out prospective cases, testing the relevance of the theoretical framework and developing a deeper understanding of Finnish radical innovations. The experts chosen were Kari Sipilä, executive director of the Foundation for Finnish Inventions and Kari Kankaala, director of Technology Transfer at Sitra (The Finnish National Fund for Research and Development). Both of them had been actively writing and speaking in public about Finnish innovations.

An executive survey was conducted in order to find out about more radical Finnish innovations. The need for this was implicitly acknowledged in the interviews, in which the experts often referred to certain industries in which there "might be" radical innovations. The survey was targeted to a carefully selected small number of influential people in Finnish business life who were known to have both long experience and the broad insight required for identifying radical innovations. Members of the Board of the Confederation of Finnish Industry and

Employers, which consisted of 40 influential executives from different branches, seemed to fulfill that criterion. It was decided to send the survey to the Board members, but since the researcher was not interested in their official statements, it was strongly emphasized that *their opinions as executives, not as representatives, of the Board were sought.*

Because these executives were assumed to be extremely busy, the survey was designed with speed and ease of response in mind. The executives were approached personally by e-mail describing the research and including a link to the Internet page on which they could very easily answer the questions and mail their response back (see Figure 5.1).

The overall number of replies was 23 (57.5 percent). Given the innovations mentioned by the expert interviewees, this was considered adequate since *the aim was to form a sufficient basis for further case selection, not to map all Finnish radical innovations.* The innovations mentioned by both the experts and the executives were then listed (a total of 51 different innovations).

First, one unclear answer was excluded because the researcher could not decipher what innovation was referred to. Second, the following innovations that were social rather than commercial were screened out:[4]

- the maternity package (a package containing mainly child-care items given gratis to new mothers);
- the national forest inventory (an inventory system of forest-resource information);

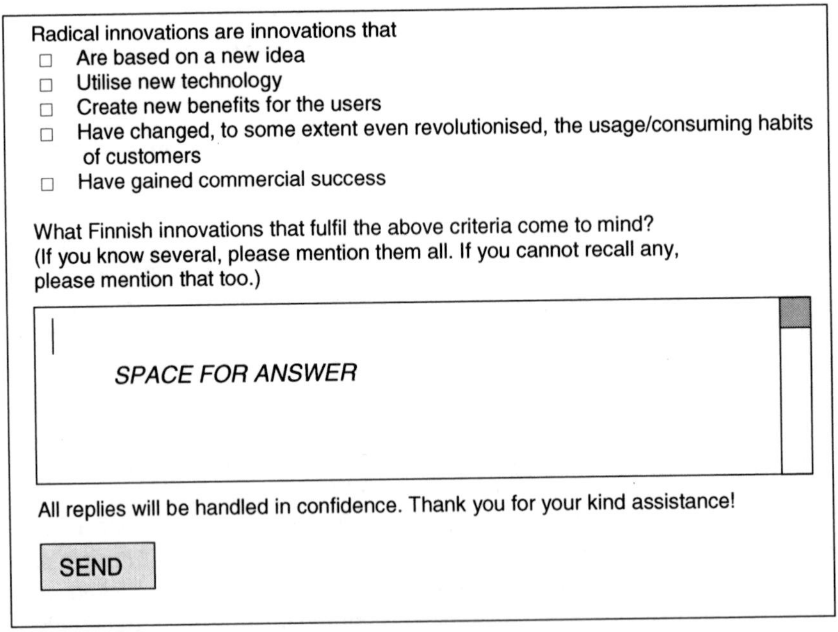

Radical innovations are innovations that

- ☐ Are based on a new idea
- ☐ Utilise new technology
- ☐ Create new benefits for the users
- ☐ Have changed, to some extent even revolutionised, the usage/consuming habits of customers
- ☐ Have gained commercial success

What Finnish innovations that fulfil the above criteria come to mind?
(If you know several, please mention them all. If you cannot recall any, please mention that too.)

SPACE FOR ANSWER

All replies will be handled in confidence. Thank you for your kind assistance!

SEND

Figure 5.1 Internet page for the executive survey.

- the draining cupboard for dishes (a cupboard in which dishes drain, thus obviating the need for tea towels).

Third, the following innovations that resembled a process or technological innovation, but could not be clearly associated with any particular commercial product or service, were also eliminated:

- the AIV silage system – the manufacturing method through which the acidity of silage is adjusted to slightly under pH 4;
- the utilization of Al/Cu foil in power transformers and reactors;
- Flash smelting – a copper-smelting process that combines the conventional operations of roasting, smelting and partial converting into one process;
- the GSM system – the Global System for Mobile Telecommunication;
- the NMT system – the Nordic Mobile Telephone System;
- text messaging in mobile telephones;
- the Linux operating system;
- the utilization of automatic data processing in the graphics industry;
- MC technology in pulp manufacturing;
- adding selenium to fertilizers;
- the utilization of urea phosphate in drip irrigation;
- pumping high densities in pulp processing;
- xylitol production technology.

Fourth, innovations that were launched before 1990 were screened out in order to guarantee that the prospective interviewees would still remember the development process (cf. Collins and Bloom 1991: 28). This applied to the following innovations:

- the Abloy lock, launched in 1918;
- biodegradable surgical nails and screws, launched in 1986;
- the electric sauna heater, launched in 1940;
- Fiskars scissors, launched in 1967;
- Harvester, launched in 1975;
- the Humicap sensor, launched in 1973;
- KERTO-laminated veneer lumber, launched in 1975;
- Radiosondes, launched in 1936;
- the Rapala lure, launched in 1949;
- silicon Capacitive Inertial Sensors, launched in 1984;
- the Suomi Submachine Gun, launched in 1922;
- the Tunturi Pulser – a heart-rate monitor for retail, launched in 1978;
- walk-through metal detectors, launched in 1971;
- Lokari mudguards – interior mudguards for automobiles, launched in 1964.

After this screening, there were 20 innovations left that had been launched in 1990 or later and which were acknowledged to be radical Finnish innovations by

Table 5.1 The criteria for radical innovations

Criteria	Definition
Finnish	Introduced by a firm that was *headquartered in Finland* at that time (cf. Chandy and Tellis 2000: 7)
Based on a new idea	Based on an idea that is *perceived as new by the potential adopter* of the product or service (Rogers 1983: 11)
Commercial success	Has been *adopted by the main market* (cf. Moore 1999a: 6–7)
New benefits	Provides *substantially higher customer benefits* relative to existing products (Chandy and Tellis 2000: 6)
Change in behavior	*Substantial change* in customer *consumption patterns* necessitated by adoption (cf. Lawton and Parasuraman 1980: 20; cf. Veryzer 1998b: 137).
New technology	Incorporates *substantially different technology*[a] or a *substantially different combination of old technologies* than existing products targeted at satisfying the same need (Chandy and Tellis 1998: 476; cf. Adner and Levinthal 2002: 55–56).

Note
a Technology can be defined as "*the practical knowledge, know-how, skills, and artifacts that can be used to develop a new product/service and/or new production/delivery system*" (Burgelman and Maidique 1988: 32).

the respondents. The researcher then evaluated these innovations in light of how they seemed to fulfill the criteria for radical innovations based on those utilized in previous studies. The criteria are summarized in Table 5.1.

Of the above criteria, the most difficult to operationalize were commercial success and behavioral change. The commercial success of an innovation has often been measured as the degree to which it brings in the expected revenues (e.g. Adams *et al.* 1998: 422; Mullins and Sutherland 1998: 226; Song and Montoya-Weiss 1998: 129). However, since it was not possible at this stage to ask this directly in all cases, the researcher had to find a measure for which the information could be found in public sources. Hence, customer acceptance was taken as a key determinant of the commercial success of an innovation (cf. Griffin and Page 1993: 297). Customer acceptance has a critical role in innovation diffusion, and an innovation that has been adopted by the main market, i.e. that has crossed the chasm between the early and the mainstream market, could be regarded as commercially successful (cf. Millier 1999: 9; Moore 1999a: 6–7). Adoption by the main market was hence regarded as a criterion, even though it was acknowledged that this could create difficulties in finding appropriate cases: adoption may take several years, or even decades (Golder and Tellis 1997: 258), and as mentioned above, innovations that were launched before the 1990 were screened out because of the difficulties associated with finding relevant data about earlier development processes.

Since the study deals with customer-related proactiveness, the "Change in behavior" criterion was adopted in referring to the most important aspect in creating a market for a new innovation, the changes in behavior that are necessary for customers to adopt the product (cf. Lawton and Parasuraman 1980: 20; Veryzer 1998b: 137). Thus, change was not understood as a macro-level phenomenon implying significant changes in the whole industry (see Song and Montoya-Weiss 1998: 126).

Table 5.2 lists the radical innovations mentioned by the executives and the experts and evaluated by the researcher in terms of how they seem to fulfill the relevant criteria. At this stage, the information about the innovations was still being collected from public sources. The most important secondary references utilized are given in Appendix 4. The refereed discussions and e-mail correspondence are listed in Appendix 2.

Thirteen innovations seemed clearly to meet the criteria set for the case selection. The potential cases were then classified on the basis of their innovative characteristics, in other words whether the core of the innovation was the product, the system or the service and whether the target market comprised other businesses or consumers (Figure 5.2). This classification emerged rather naturally as the researcher started to sort out the cases that were left for the final case selection.

Multiple cases were selected in order to obtain a more complete picture of the innovation phenomenon and more extensive understanding of how it operates in different companies and in different situations. There are no precise guidelines regarding the number of cases to be included in any one study. The range presented by various researchers tends to vary between 2 and 15 (Perry 1998: 793–794). Eisenhardt (1989: 545), for instance, suggests that four to ten cases

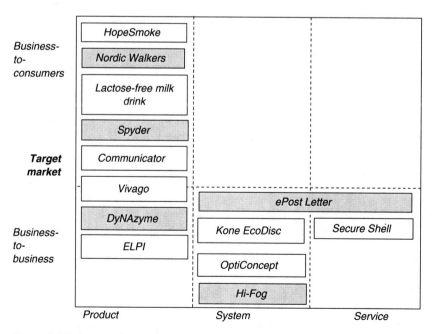

Figure 5.2 Selection of the five cases.

Table 5.2 Evaluation of radical Finnish innovations

Named radical innovations		Named by		Criteria					
				for Finnish innovation			for radicalness		
Innovation (main sources utilized in the evaluation)	Firm that launched the innovation (year of the launch)	Executive	Expert	Finnish	New idea	Commercial success (estimated in early 2002)	New benefit	Change in behavior	New technology
Benecol cholesterol-reducing ingredient	RaisioGroup (1995)	yes	yes	yes	yes		yes	yes	yes
Communicator	Nokia (1996)	yes		yes	yes	yes	yes	yes	yes
Azipod contra-rotating propulsor	ABB and Kvaerner Masa-Yards (1990)	yes			yes	yes	yes	yes	yes
ELPI, electrical low pressure impactor	Dekati Ltd (1994)		yes	yes	yes	yes	yes	yes	yes
ePost Letter	Finland Post Corporation (1992)	yes		yes	yes	yes	yes	yes	yes
Handy Axes	Fiskars (1991)	yes		yes	yes	yes	yes		yes
Hi-Fog, water-mist, fire-extinguishing system	Marioff Corporation Oy (1991)		yes	yes	yes	yes	yes	yes	yes
Imaging phone	Nokia (2002)		yes	yes	yes		yes	yes	yes
KONE EcoDisc® hoisting machine	KONE Corporation (1996)	yes	yes	yes	yes	yes	yes	yes	yes

Product	Company							
Lactose-free milk	Valio (2001)	yes			yes	yes	yes	yes
Left-foot, mass-customized men's shoes	Left Foot Company (Pomarfin) (2000)	yes		yes	yes	yes	yes	yes
Magnetically controlled shape memory materials	AdaptaMat (2001)		yes	yes	yes	yes	yes	yes
Nordic Walker poles	Exel (1997)		yes	yes	yes	yes	yes	yes
OptiConcept paper machine	Valmet-Rauma (1998)	yes	yes	yes	yes	yes	yes	yes
PARKIT Parking payment system	Payway Oy (2002)		yes	yes		yes	yes	yes
Secure Shell Internet encryption systems	SSH Communications Security (1997)		yes	yes	yes	yes	yes	yes
HopeSmoke smoker bag for oven or grill	Oy HopeFinlandia Ltd. (1993)		yes	yes	yes	yes[a]	yes	yes
Spyder wristop diving computer	Suunto (1997)	yes		yes	yes	yes	yes	yes
DyNAzyme thermostable DNA polymerase	Finnzymes (1991)		yes	yes	yes	yes	yes	yes
Vivago WristCare System	IST International Security Technology Oy (1998)		yes	yes	yes	yes	yes	yes

Note

a The product reached market quickly and showed profits. The later bankruptcy of the firm seems to have been related to contract disputes, not to the product itself.

tend to work well. In this study, the number of cases was determined along the process, when certain disparities and certain congruence had become evident (cf. Glaser and Strauss 1967: 61–62). In the end, *five* cases seemed to provide enough information and were still manageable.

In order to facilitate the modification of the a priori framework to suit different types of radical innovations, a heterogeneous group of cases was selected for further analysis. The selection was not motivated by aspirations toward generalizability, since this was considered to be beyond the scope of this study with its restricted number of cases. On the contrary, different cases were included in order to achieve a more holistic understanding of the phenomenon itself (cf. Van de Ven and Poole 1990: 316).

The cases were thus selected so as *to deliberately secure variation in the nature of the market and the commodity*. Past studies suggest that there may not be considerable differences in the development processes of different kinds of innovations. Sirilli and Evangelista (1998) compared service- and product-innovation processes and concluded that they exhibited more similarities than differences. Both rely on a wide range of innovation sources, often have the same aims (improving quality, increasing market share and reducing production costs), face the same obstacles along the way (related to high cost and risk) and use various sources of information, although internal sources are regarded as a very important technological information source in both service and product innovation (Sirilli and Evangelista 1998). Furthermore, it has been argued that services are becoming increasingly technology-intensive, and manufacturing is becoming more service-like, which means that the traditional demarcations between services and products may be eroding (Miles 1994: 252).

Although Holt *et al.* (1984: 37) suggested that the type of market may influence need assessment in innovation development, Millier (1999: 31) argued that in the high-technology field, in which most radical innovations are created, the development of b-to-b products and services is rather similar, since both are mostly intangible, and therefore "customers never know beforehand what they're buying" (see also Tidd *et al.* 1997: 178).

The five selected cases were: Hi-Fog (a registered brand of the Marioff Corporation), Nordic Walkers (Nordic Walker is a registered brand of Exel Sports Ltd.), Spyder (developed by Suunto Ltd.), DyNAzyme (a registered brand of Finnzymes Ltd.) and ePost Letter (developed by Finland Post Corporation).

Hi-Fog is a water-mist fire-extinguishing system developed for ships, but later also sold to underground transport systems, industrial buildings, hotels and museums, for instance. Nordic Walkers are specially designed poles for fitness walking. Spyder is a wrist-top diving computer. DyNAzyme is a thermostable DNA polymerase utilized in gene technology. ePost Letter is an automated letter-mailing system that combines electronic communication and traditional mail: it was sold as a service to businesses and as a system to various national postal operators.

Thus, this study is clearly not limited to a certain industry, even though such a limitation certainly would have facilitated the research because then the outer context of the firms would have been rather similar. On the other hand, as Verganti (1999) suggests, proactiveness seems to be rather independent of the industry. Furthermore, according to Veryzer (1998b), the customer-research efforts of firms developing radical innovations seem to be quite similar across industries, and the limitation to a certain industry was not considered a problem in this study.

In sum, the selected five cases have certain commonalities: they are all successful Finnish radical innovations that were launched during the 1990s. Moreover, they form a heterogeneous group ranging from product to system and service and from business-to-business to consumer-market goods.

5.1.3 Data collection

The study is based on longitudinal historical analysis, which posed challenges in terms of the data collection. Access to research objects is often problematic in historical studies, since the time has passed. Study of the past is always dependent on documents, artifacts and other people's recollections. Hence, it can never be experienced again, and the researcher is forced to remain far removed from the studied objects (Halinen and Törnroos 1995: 512; Parker 1997: 133). In this case, the concentration on recently launched innovations allowed the creation of new data in the interviews. Thus, *interviews* were used as the key method of data collection.

In order to form a complete picture of the innovation development processes and to compare each process with the initial framework, the researcher had to balance the need for getting an unprompted story with the need to get answers to specific questions related to the purpose of the research (cf. Mason *et al.* 1997: 314). This prompted the choice of *semi-structured interviews* as the main data-collection method.

The selected themes were based on the initial framework and modified following the expert interviews and the pilot study. Since it was found out in the pilot study that different interviewees tend to understand the innovation development phases in different ways, frequent checks were made to make sure that the interviewer and the interviewee were speaking about the same stage. Further, for the sake of clarification, both parties together drew the chronology of the development stages at the beginning of each interview.

It has been suggested by various researchers (e.g. Collins and Bloom 1991: 29; Yin 1989: 89) that interviews should be used in conjunction with written data in order to amplify, confirm and modify the record. In fact, each case study began with a search for information about the innovation in public sources, mainly magazines, newspapers and the Internet (see Appendix 3). Since all the innovations being studied were largely known about, the public information was abundant. It is acknowledged (Mason *et al.* 1997: 312) that public sources are important for establishing the authentic timeline, which in this case served as background information when preparing for the interviews.

The public sources were also utilized as data triangulation; information obtained from them was compared to the data gathered through the interviews. Triangulation facilitates the combination of various data-collection methods, and whereas interviews may provide depth and personal feeling, secondary sources may provide factual information (Forster 1994: 148; Pettigrew 1990: 277).

Although public sources were valuable in this case, the information obtained from them was scrutinized carefully. It was acknowledged that it was produced for different purposes, and some contained more gloss than others. Furthermore, as Mason *et al.* (1997: 313) also noted, these kinds of sources mentioned very little about the trials and tribulations, which were then revealed in the interviews. On the other hand, newspaper and magazine articles have also been rated as high in authenticity because the time between the event and its reporting is short (Golder 2000a: 160).

The number of interviewees for each case varied from two to five. This variation was natural since, in some cases, a certain person was involved in all of the phases from the idea to the launch stage, whereas in others it was not possible to form a complete picture without talking to several interviewees. A total of 14 people were interviewed. Two interviews were conducted in pairs following the suggestion of the interviewees themselves. All the interviews were conducted in the case firms' premises, and they lasted from one to three-and-a-half hours and were tape-recorded and transcribed in order to increase the trustworthiness of the research.

In order to eliminate possible errors, the interview data related to certain cases were crosschecked and, in case of contradictory evidence, the interviewees were contacted by telephone or e-mail to clarify the points in dispute. In order to correct factual errors and to check for the inadvertent release of commercially sensitive information, the researcher gave all the interviewees the opportunity to review the case description before its publication (cf. Odendahl and Shaw 2002: 313–314). The review process did not cause any major changes in the case descriptions but added some clarifications and a few corrections of factual errors.

5.1.4 *Data analysis*

According to Leonard-Barton (1990: 263), the analysis of interview data requires a high tolerance of initial ambiguity, as "one iterates toward clarity." The empirical data for each case in this study were abundant. Several researchers (e.g. Halinen and Törnroos 1995: 506; Savitt 1980: 53; Van de Ven 1988: 333) have suggested that analyzing longitudinal qualitative data ought to begin by putting them in chronological order. This was a rather natural step here too, and the preliminary classification thus took the form of a *chronological listing* of events. The time line helped in the organization of the large amount of empirical data.

Second, in the interpretation phase, the data were organized thematically according to the a priori framework. It has been suggested that this kind of analytical chronology helps to clarify patterns in the data and to establish sequences across levels of analysis (Pettigrew 1997: 346; Yin 1981: 64). The theoretical framework of the study was linked to the cases via *pattern-matching* logic (Pauwels and Matthyssens 2004: 130; cf. Yin 1989: 109–113). Pattern matching facilitates the relation of several pieces of information from the same case to the initial framework and helps to concentrate attention on the key factors and the purposes to be fulfilled. On the other hand, it carries the risk that certain new themes arising from the empirical data may remain unnoticed. The researcher kept this risk in mind and made every effort during the writing of the case descriptions not to neglect relevant issues that did not "fit into the initial frame-work."

The case descriptions were presented as narratives that were built on the structure of the a priori framework. According to Mason *et al.* (1997: 317), "The penultimate step in an historical study is to tell the story." In this study, the stories were told in the form of case descriptions. Each one is the researcher's interpretation of how the innovation was developed. It is a narrative that enables the researcher to communicate a rich understanding of events and link them together in a way that is not possible with simple chronologies (Golder 2000a: 161).

The interviewee citations were included in their translated form in the case descriptions.[5] They were utilized to justify the interpretations and to enliven the descriptions. As each process of radical innovation development was such a unique episode in the particular firm, and also widely known in the industry, it would have been impossible to protect the identity of the interviewees. There-fore, no attempt was made to maintain their anonymity in the case descriptions or in the citations, which were identified by means of the interviewees' job titles. All the interviewees approved of this practice.

Each case is presented separately, and then they are compared with each other in order to find patterns of similarities. *The case comparison* was facili-tated in that the same structure was used for each description. Finally, the main theoretical and empirical findings were linked to wider bodies of literature.

In sum, the analysis of the data was a continuous process that required repeated reading of the interview text files, the notes and the secondary data. It involved the researcher returning to the theoretical literature and to the inter-viewees with additional questions. This constant comparison between theory and empirical reality resulted in the creation of a modified framework.

5.2 DyNAzyme

5.2.1 DyNAzyme as a radical innovation

Gene technology, i.e. the technology used to manipulate the genetic composition of an organism, was under development in the second half of the twentieth century. One of the major discoveries pushing this technology forward was made in 1968, when the first restriction enzyme was isolated and characterized. Restriction enzymes are enzymes that protect the organism from invasion by foreign DNA. They are usually isolated from bacteria, where they defend the host against foreign DNA. They scan a DNA molecule and look for a particular sequence, usually of four to six nucleotides. Once they find this recognition sequence, they stop and cut the strands. Hence, they cut DNA into manageable and reproducible pieces. Furthermore, since enzymes are able to cut DNA at specific sites, and these cut ends match up with other cut ends, it is possible to cut both donor and host DNA and then mix the two and get them to rejoin. Consequently, a gene can be transferred from one organism to another. Restriction enzymes thus enable the creation of entirely new kinds of DNA molecules, as well as the manipulation of the genes located on them. Enzymes that catalyze the formation of DNA and RNA from an existing strand of either are called polymerases.

The polymerase chain reaction (PCR) technique, the subject of the Nobel Prize in chemistry in 1993, was invented in 1983 by Mullis. It is utilized for rapid in vitro amplification of a specific fragment of genomic DNA or RNA. By enabling the extremely fast copying of genes, it revolutionized the way in which molecular biology was conducted. It also made experimentation with DNA much more affordable. The PCR technique needs high temperatures, which places considerable demands on polymerases, which are proteins and proteins denature as the temperature increases. Consequently, there was a need to develop thermostable polymerases that retain activity at elevated temperatures and can thus be used repeatedly for DNA amplification.

These polymerases were sought and purified from the hot springs bacterium. *Thermus aquaticus* (Taq) DNA was the first commercialized thermostable polymerase. It was introduced in 1987 and was soon followed by other polymerases, one of which was DyNAzyme (Figure 5.3). The Finnish firm Finnzymes Ltd launched DyNAzyme in 1991. It was isolated and purified from a strain of *Thermus brockianus*, a Finnzymes proprietary bacterial strain.

DyNAzyme was a *commercial success*. There was a strong emerging demand for these kinds of enzymes by the research community and in biotechnology. The turnover of Finnzymes grew very rapidly following the launch of the polymerase in the early 1990s (Figure 5.4). Since then, DyNAzyme has been the basis of a whole line of PCR enzymes and kits that Finnzymes launched. It also

Figure 5.3 The DyNAzyme product (picture of tubes by Eeva Sumiloff).

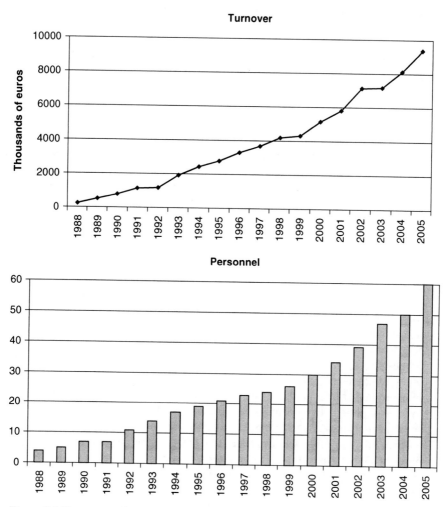

Figure 5.4 Turnover and personnel of Finnzymes 1988–2002 (Holm, e-mail July 1, 2004; Voitto+ 2007).

produced image-related benefits that contributed to the firm's success: "For image reasons it was a really big thing for us in Finland, which obviously also increased our credibility in the eyes of the customers, which in turn affected our total sales, too" (Export Manager).

The application areas for DyNAzyme were vast, including gene technology, the diagnosis of genetic diseases and human DNA profiling. It was recognized as a significant Finnish innovation in 1994, when Finnzymes was awarded first prize in the Innofinland contest. In terms of *technological novelty*, it was a new enzyme that inherently had different features than existing enzymes. Since these features are closely reflected in the benefits, both are discussed next.

DyNAzyme brought significant *benefits* to the adopters when it was launched. The thermostability of DNA polymerases is extremely important when they are used in amplification procedures. Increased thermostability reduces activity loss during thermocycling and allows higher denaturation temperatures for the amplification or sequencing of difficult templates. When DyNAzyme was launched, its thermal stability was better than that in other polymerases: it has a half-life of three hours at 96°C.[6]

High fidelity is also important in PCR applications such as cloning. DyNAzyme lacked 3′→5′ proofreading activity,[7] which made it easy to use but increased the possibility of unintended secondary mutations. Despite this short-coming, however, its error rate was half that of Taq polymerase. DyNAzyme also contained a 5′→3′ exonuclease activity that is involved with excision repair within a cell: it enables the polymerase to replace nucleotides in the growing strand of DNA by means of nick translation.

These benefits influenced the *behavior* of the adopters. First, because of its broad working range, DyNAzyme was easy to use and optimize in different PCR applications. Second, better thermal stability led to better activity and stability at high temperatures. Better thermal stability also increased the utilization time of the enzyme and, consequently, the user needed a smaller amount. Hence, the costs came down. Better thermal stability also led to better stability upon storage and transport.

However, it is worth pointing out that it is impossible to create an optimal thermostable polymerase that would be the best solution for every researcher. A variety of elements (polymerase, primers, temperatures, times, template DNAs) are intermingled in a number of ways in PCR, and researchers therefore need to have many options from which to choose to find the most suitable polymerase for each experiment. DyNAzyme increased this choice.

There are also other features that have affected the development and marketing of DyNAzyme. First, it is a *delicate* product that needs to be treated, stored and transported carefully and at a specific temperature (−20°C). Second, it is an extremely *small but expensive* product, which is sold in very small quantities. For example, in the early 1990s, one kilogram of DyNAzyme cost around 200 milliard Finnish marks (i.e. over 33 milliard euros).[8] Third, it is a business-to-business product that could be more specifically described as *an expert-to-expert product*. The main market consists of research laboratories in universities, hospitals and pharmaceutical firms, and there are a variety of application areas ranging from genetics to microbiology and plant bio-technology.

5.2.2 Finnzymes Ltd. – the developer of DyNAzyme

Pekka Mattila and Kari Pitkänen were students on a biochemistry course at Helsinki University of Technology in 1984. The lectures about DNA modification interested them, and they were inspired by the fact that no Finnish firm produced restriction enzymes. Consequently, the following year, as part of the Plant

Design course, they made a business plan for a restriction enzyme factory. Although they received a poor grade, they were convinced that their plan was good and offered it to a big pharmaceutical firm. The firm did not consider the idea to be feasible. The two then applied to Tekes (the National Technology Agency of Finland) for funding for a small project concerning the planning of a restriction enzyme factory. Their application was approved. Thus, their obstinacy and their trust in their own idea paid off.

In order to learn as much as possible about restriction enzymes, Mattila and Pitkänen contacted the world's best laboratories in the field, including New England Biolabs (NEB), Boehringer Mannheim (currently Roche Diagnostics) Fermentas, the University of Bristol in England and the University of Leiden in the Netherlands. All these laboratories were cooperative, but the most important contact was formed with NEB, who invited them to visit them and to learn about the handling of restriction enzymes there.

Although some aspects of the collaboration were secret, most of it was very open. After all, it was thought that various researchers in the field had already anticipated most of the emerging issues in biotechnology. Mattila and Pitkänen visited NEB and other laboratories several times. It was surprisingly easy to gain access to these places:

> The first time I went to the States I was told that "you're never going to see anything there, at most you might get to be in the meeting room". [...] And when I went to Biolabs for the first time they were immediately like "so, what would you like to do?" It was really open as I'd never been there before and I had access to everything, right from the start.
>
> (CEO)

Hence, the techniques were learnt in NEB. More funding was also received from Tekes. In 1986, Pekka Mattila, Kari Pitkänen and Jens Stormbom, an old friend of Mattila who was interested in investing in biotechnology, decided to set up a firm. Thus, Finnzymes was founded as a non-operating company.

In the summer of 1987, the president of NEB suggested that they could form a joint venture with Finnzymes, thus the dormant firm was activated. The shares were distributed among five people: Mattila (the current CEO), Pitkänen (the current marketing director), Tenkanen (the current R&D director), Stormbom (the current Chair of the Board) and NEB.

The activities in Finnzymes have been twofold from the very start. First, it has its own research and development, the aim of which is to provide new molecular biology products of extreme purity to the scientific community. Currently, the main products are DNA and RNA polymerases and transposon-based[9] tools for functional and structural analyses of genes and proteins. Second, the company distributes a wide range of products used in molecular biology. Although acting as a distributor may have taken attention away from its own research and development, it proved to be extremely important, especially in the beginning:

It was a good thing for us in a lot of ways. We got cash flow to the company at the beginning when we didn't have any of our own products. Another significant thing was that we learned how to sell the products, because we were just a bunch of broke students at Helsinki University of Technology. [...] I didn't have any of this venture money to hand then. There weren't that many funds like the Finnish National Fund for Research and Development or the current Biofunds.

(CEO)

Furthermore, having the same customers for both the imported and their own products brought in synergy benefits. Close contacts with customers were utilized in the research, and knowledge acquired from the customers' needs was incorporated into the development of new products. Moreover, Finnzymes had close contacts with distributors in other countries, and it was able to use this network when distributing its own products. Consequently, distribution and its own development seemed to support each other.

In the late 1980s, Finnzymes began its own production of restriction enzymes together with The Technical Research Centre of Finland (VTT) and NEB. Research visits to NEB played an important role in the setting up of its own production. The turnover in 1991 was around 1.1 million euros, and it employed seven people. It had a very *organic* organizational structure at that stage. The flow of information in the firm was very free and, due to its small size, there was no clear division of labor between product development and marketing, so that everyone participated in both. The emphasis has been on getting things done rather than on formal bureaucracy, and the organization has been rather informal: "We've always had a very flat organization. We haven't had a clear, structured organization. Sometimes that's good and sometimes that's bad. Sometimes it feels like everyone's interfering with everything" (CEO). In the beginning, the decision-making was more concentrated in the hands of the few key people. These people were biochemists and they learned about the management of the firm by doing it:

We haven't really received any special training on how to manage a company. We've learned to do the right things. [...] Well, we've learned a lot of business stuff the hard way. Let's just say that we've had to learn the basics of business, such as marketing, from the beginning.

(CEO)

Both Mattila and Pitkänen seem to have *proactive personality* characteristics, such as extraversion and conscientiousness, which could be seen in the way they set their goal to develop a thermostable polymerase and managed to carry that plan through by establishing contacts with various cooperation partners. Mattila also highlighted the need for activity in the interview: "I do think that those who can, do, and only by doing you'll learn. That's the most important thing. Not that you can read books but that you get on with it, that's the most essential thing" (CEO). The proactiveness was not restricted to the innovation development but also extended to the customers. This was evident in their attempts to anticipate what kind of polymerase the customers would need and, in particular, in their persistence in convincing them that the DyNAzyme would fulfill these needs.

The *leadership* in Finnzymes seems to have concentrated strongly on product development. Product development capabilities were built up primarily by recruiting skilled scientists, and marketing capabilities were not particularly accentuated. The training of employees was related to their tasks more than to their individual preferences. Hence, the training was often directed toward enhancing marketing skills.

The management facilitated new-product development by allocating as many resources as they could. From the very beginning, the aim was to generate growth through new innovative products, and this aim has been communicated throughout the firm. The organizational culture of Finnzymes seems to have reflected a strong intensity and the crystallization of proactiveness with regard to product development and scientific research. Brainwork and expertise were emphasized throughout:

> The whole organization is completely aware that we always need new products. I don't think that's unclear to anyone. If we don't have them, we think it's a bad thing. It's a bad thing for everyone, no matter what their role is in this company.
>
> (Export Manager)

Product development was also dominant in the stories about incidents during the research projects. Although it was accentuated in the organizational culture, the importance of customers was also acknowledged:

> We're small enough so that everyone understands that we don't automatically just work if we don't have customers. I think it's sunk in quite well. [...] I believe that the researchers and the marketing people feel some sort of pride if we're aiming to create a product that is internationally interesting, possibly the best, which is just great. Just great as such. But of course its financial significance ... everyone is extremely pleased by that.
>
> (Export Manager)

The creation of a product that achieved worldwide success was strongly tied to the environmental circumstances. The features of the environment are discussed in the next section.

5.2.3 Environmental turbulence

Gene technology underwent considerable changes in the 1980s. Developments in computer technology, software and computer networks created new opportunities for genetic research and genetic data management, and genetic technologies such as PCR were emerging. PCR was gaining a foothold in laboratories in the late 1980s, and this encouraged scientists to optimize this technique. The enzymes used were put under scrutiny, and there was a growing need to discover new and better polymerases and to create optimal mixtures for certain applications. Better thermostable polymerases were required to turn PCR into an efficient technique.

Furthermore, PCR opened up new avenues for research, which further generated new biotechnological inventions. The field of genetic research was very dynamic in the 1980s, and research groups were competing to be the first to publish new findings. Thus, the environmental turbulence in the biotechnology industry was enhanced by the considerable rapid changes, which increased the dynamism of the environment (see Figure 5.5).

Since the changes were abundant and often interrelated, one invention leading to others, the *complexity* of the environment could be considered high. However, although the changes were continuously reshaping the industry, the general trends seem to have been at least partially anticipated by the researchers working in the field. For instance, the founders of Finnzymes were able to anticipate the future growth of PCR technology, presuming that the demand for better thermostable polymerases was going to grow. This therefore sparked their interest in developing the polymerases. Hence, it could be argued that some future changes were to be predicted: at least some were linear, and the researchers knew where they were heading, even though biotechnology was such a nascent and evolving field that there were no data concerning its development paths. Predictability thus decreased the environmental turbulence in the biotechnology industry.

The turbulence in the industry is also reflected in the attempts to regulate it. New inventions in gene technology have resulted in an increasing amount of regulation. Already in 1975, at the Asilomar Conference scientists recognized the need for regulation, and they developed experimental guidelines for laboratory research using genetic engineering. Nowadays, it is one of the most tightly

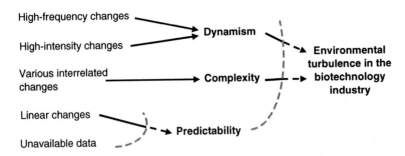

Figure 5.5 Factors contributing to the environmental turbulence in the biotechnology industry (the degree of unity of the arrows illustrates the degree of influence of the factors).

regulated industries. It is supervised in Finland through the Gene Technology Act that was adopted in 1995. The Act aims to prevent and avert any harm that the use of genetically modified organisms may entail to human health, animals, property and the environment. It also promotes the safe use of gene technology and its development in an ethical way and is in line with the corresponding European Union (EU) directives. Compliance with the Gene Technology Act is monitored by the Ministry of Social Affairs and Health. The CEO of Finnzymes felt that they had been able to influence the content of the Act, since they were invited to comment on the regulation proposals when it was still under development.

5.2.4 Idea generation

As a distributor, Finnzymes had knowledge of PCR equipment and DNA polymerases, which it was basically interested in producing. Mattila and Pitkänen had learned how to isolate, characterize and purify restriction enzymes in their earlier research visit to NEB, but they lacked knowledge of how to control the quality during the production process. This knowledge was again acquired during a research visit in 1988, when Finnzymes sent one of its chemists to work in NEB for eight months. The aim of the visit was to learn more about enzyme production.

A meeting to plan a Nordic biotechnology program took place in Norway in 1989. The growing importance of PCR technology was one of the themes that was raised in the meeting. The CEO of Finnzymes met an Icelandic researcher, Jakob Kristjansson, a specialist in thermophilic bacteria. In unofficial discussions, they started to contemplate the possibility of jointly searching for and purifying DNA polymerases from bacteria existing in the hot springs in the geothermal fields of Iceland.

Finnzymes had the necessary equipment for analyzing DNA polymerases, and it had already tested one thermostable polymerase, which was not suitable for further analysis. Kristjansson's research team (IceTec) had long experience in researching bacteria in Icelandic hot springs, and cooperation seemed potentially beneficial for both. They assumed that the springs might hold better polymerases than the *Thermus aquaticus* (Taq), which was the market leader.

The criterion they set was that the new polymerase should have better thermal stability and better proofreading capability than the polymerases currently in use. At that time they strongly believed that if they could manage to find that kind of polymerase and were able to fabricate it, it would quickly create a market for itself. This belief was supported by the *anticipation* of future market opportunities:

> It was clear to me ... well, not exactly clear what size the market was, but everyone realized that it's got to be an important technology so that gene transfer will advance. In a way, everyone understood that this has really got

to have unlimited potential uses. In that sense it has the whole market to itself.

(CEO)

Thus, due to the nature of the gene-technology industry, the geographic scope could be called *geocentric*. Although certain regions, such as Northern America, Europe and Japan, were considered more important than others, it was rather clear that the market for thermostable polymerases would extend beyond them.

However, anticipation of market opportunities was not sufficient to trigger the search for DNA polymerases. In order to realize the idea, Finnzymes and IceTec needed external finance, and they applied for funding from the Nordisk Industrie Fond. Their application was approved in 1989, and the research project could be initiated. This was the beginning of the development stage.

5.2.5 Development

The development stage began in 1989 with the screening of the bacteria. The Icelandic research team collected bacteria from the local hot springs and did the preliminary screening in order to find out whether they contained DNA polymerases. Around 300 of the bacteria that were collected were screened, and those containing DNA polymerases were sent to Finnzymes for further analysis. Finnzymes continued to study these polymerases in order to find one that was better than the Taq.

In the late 1980s, there were around 20 firms in the world that were developing thermostable polymerases, most of which operated in the United States. Finnzymes was the only one that was located in Northern Europe. The emergence of new competitors was restricted by the lack of funding for this kind of development project, but in any case, Finnzymes was not particularly worried about the competition. The following statement describes their attitude quite aptly:

We did know we weren't the only ones. It's obvious that these PCRs are always competitive. But on the other hand we didn't really think about that sort of risk there. If you find a better enzyme than the other people have, then so what?

(CEO)

Consequently, the firm openly admitted in interviews in newspapers and magazines that they were starting a project that aimed to produce thermostable DNA polymerase.

During the process of analyzing the enzymes, Finnzymes came up with a DNA polymerase that seemed to fulfill all the criteria set at the beginning. It was isolated from a strain of *Thermus brockianus* bacterium and purified of contaminating endonucleases and exonucleases at the beginning of 1990. Purifying an enzyme was a demanding process, which included getting rid of hundreds of noxious enzymes in order to produce one good one and testing various applications. After purification, the enzyme was characterized, which means the sites at which it cuts DNA were specified. Market research would have been impossible at that stage, and they had to rely on their intuition in terms of what would be important.

Before the development of the prototype, Finnzymes relied on its own experience of restriction enzymes and its knowledge of the end users gained through selling NEB products to the Scandinavian market. Consequently, the company was able to *anticipate* what kind of polymerase the customers would need. Although its direct customers were from one region, it was thought that customers of polymerases would be similar and have the same kinds of needs throughout the world; thus, the anticipation was *geocentric*.

After purification, the enzyme had to be optimized, which means specifying the limits within which it functions properly. Optimizing the temperature and the template concentrations takes place through trial and error. In the case of DyNAzyme it took over six months, so the product was ready to be launched at the end of 1990. At this stage, i.e. after the prototype development and during the optimization phase, there was still no chance of doing market research:

> Usually you notice something in the research material that you think might be significant, make a product out of it, launch it and then customers either use it or don't use it. It's like bringing out tools and telling everyone what to do with them. But defining the market, that's just something shaken out of your sleeve.
>
> (CEO)

However, Finnzymes was again able to *anticipate* the needs of the end customers, because it knew the retailer market in the Nordic countries and had direct contacts with other retailers. It was also actively following the scientific discussion and emerging trends. Since the needs of the researchers in that field were considered to be global, the scope of the anticipation was *geocentric*.

The developers were extremely enthusiastic about the new enzyme and described it as their child. This *enthusiasm* spread throughout the small organization. People were working long days and during the weekends, and when it was ready they were extremely eager to create a market for it.

5.2.6 Creating the market

DyNAzyme was introduced to the market in 1991. The time was right – just as the increasing use of PCR technology was boosting the need for thermostable polymerases. Hence, the firm was able to reap first-mover advantages. Although getting onto the market quickly was considered important, it was more important to be convinced about the quality of the polymerase before the launch:

> The way we think about quality is that we only have one chance. We're such a small company that we'll never get a second chance, it has to work

straight away. That's why we're a bit overly critical when it comes to product quality when we launch something. We can't let anything leave our hands unless we're absolutely sure that we're happy with it in every way, as we can't afford compromises.

(CEO)

Quality was especially important since the customers used the polymerase in their own biotechnical trials, and they had to be sure that the results of their trials were not affected by the inferior quality of the polymerase they used. These features of biotechnical trials further increased the barriers to the adoption of a new polymerase: "If they have a certain way of conducting experiments and it works well, it's very difficult to get them to change their ways. Even if it makes sense financially or timewise, then still ... because 'this works just fine!'" (CEO). According to the CEO of Finnzymes, researchers are therefore often rather conservative when buying polymerases. The first people to adopt the new one were innovative and enthusiastic researchers:

Maybe they are the kind of people who think more for themselves, who don't just take everything out of a book. It depends a lot on what the researcher is like, that he's not a researcher just for the sake of it, but that he's got the mindset of a scientist, that he really wants to know and is curious. As a general rule I'd say that younger people are more eager to experiment.

(CEO)

The willingness to experiment was enhanced through influencing end customers both directly and via distributors.

Influencing the customers

The first step in the launch was to visit the retailers and educate them about the new polymerase. At the same time, it was important *to influence* the end users directly in various countries. These potential customers (universities and research institutes) were visited together with the local retailer. The aim of the visits was to raise *awareness* of the polymerase and to teach people how to use it, as the CEO said:

You could say that I just grabbed a bag and traveled a lot. It was pretty much about going there to teach the distributors and teaching the customers together with the distributors ... It's about meeting the end-user, the customer, and personally guiding them through it.

(CEO)

Generally, the teaching of DyNAzyme utilization was considered rather easy because the researchers were familiar with the PCR technique. Since the product was also being sold by the researchers, the sellers were on the same wavelength as the customers and were able to understand their needs.

In order to obtain information about the polymerase and its quality, customers often relied on word-of-mouth communication:

> It really is that when someone says that this is a good one then … Word of mouth really does work well! Both for the good and for the bad. The bad information spreads quickly too. And another thing is the Internet, where there are a lot of discussion forums about our field. If something works there and something doesn't, then they'll know it instantly in the discussion forums.
>
> (CEO)

Influencing the word-of-mouth communication was considered difficult. The most important task had been accomplished during the development stage, when the quality of the product was ascertained. According to the CEO, the only thing that could be done at the launch stage was to provide plenty of accurate information about the polymerase and its application areas. He acknowledged that more resources would have been needed in order to produce more application sheets, i.e. to find out and demonstrate various application areas for DyNAzyme.

Scientific conferences and biotech trade fairs were also important avenues in creating both awareness about DyNAzyme and *getting researchers to try it*: reference customers have been particularly useful in biotechnology. Foreign distributors were invited to Finland to discuss both technical and marketing issues, and distributors were provided with the opportunity to conduct laboratory trials. The main selling points have been the quality of the polymerase and the price. The superior quality of DyNAzyme made the selling of the polymerase rather easy, although competition on the enzyme market was getting harder.

Selling was supported through advertising. Advertising in international scientific magazines (e.g. *Nature, Natural Science, Biotechnics and Science*) was used to create awareness. Although the advertising was expensive, it was considered an important tool for image building:

> So far we've mostly been in this kind of bigger publication, which in a way shows you something about the status and boosting the image of the company. We do get feedback on it. I think that one target group, even if it sounds a bit silly, is also distributors. When they see that we're advertising in the magazines then it does add to the credibility of our company. As we're not a very big company, this is the way we've got to do it.
>
> (Export Manager)

Thus, advertising was considered essential. The features of DyNAzyme were often compared with those of other thermostable polymerases. At first, the advertisements and brochures were rather miscellaneous, and there was no attempt at consistency. The lack of resources also limited the planning of marketing communication.

The market for DyNAzyme in Finland was rather small, and therefore, international distribution was extremely important. This is discussed in the next section.

Establishing the international distributor network

Finnzymes sold DyNAzyme directly to its Finnish customers, but abroad it utilized an international network of distributors. This was a natural choice for a small biotechnology firm that itself acted as a Finnish distributor of the products of NEB. It had formed contacts with other foreign distributors in distributors' meetings and used this contact network in the distribution of DyNAzyme. The NEB distributors dealt with high-quality products and were therefore suitable for selling DyNAzyme. Furthermore, since DyNAzyme did not compete with NEB's own products, the utilization of its distributors did not cause problems.

Thus, it seems that NEB also played a significant role at the launch stage. In selecting the distributors, emphasis was put on their expertise and complementary product portfolio. They were rather easy to find, since it was known in the field that Finnzymes had cooperated closely with NEB, which was one of the leading firms in the industry. However, the lack of resources somewhat slowed down the international launch.

The first stage was to explore the markets of Northern Europe, and this was soon followed by expansion further into Europe. This was a natural starting point since Finnzymes knew the distributors, and DyNAzyme became established in the European market in 1995. Europe was an important target area since it accounted for around 40 percent of the world's polymerase market, another 40 percent being covered by Northern America.

Although the export sales in Europe were a success, entry into the United States was slower than expected. One reason for this was the competition that was becoming very hard there. Sales to the United States started to grow in 1997, but it has been a laborious process due to the strict requirements that the U.S. Department of Agriculture has set on imports, particularly on biotechnological products:

> They want to know where you've brought it from, how it's been done, you've got to tell them about the production, everything ... Despite everything, even if all the permits are OK, the product sometimes gets stuck in customs and goes off there or the customer doesn't want it anymore when it's been lying there for three days without anyone caring. Even if it says outside the package what temperature it should be kept at, well, that package is gone and so we make another one. No-one takes responsibility, that's slightly frustrating.
>
> (CEO)

Despite being a difficult export market, the United States is nevertheless regarded as a key target market for Finnzymes. Other export markets grew rapidly from the beginning, although international trade restrictions have limited the export of polymerases to certain countries, such as where there is an elevated risk of their usage in the development of biochemical weapons. Nowadays, Finnzymes products are sold around the world, and exports cover more than 90 percent of the turnover. The United States and Japan are the countries in which efforts are being directed at gaining a firmer foothold.

Although distributors in each country have their own style of marketing the product for the end customers, customer behavior is considered to be rather similar throughout the world:

> I've been talking directly to researchers in several places, so regardless of culture, an Asian, an American or someone else, people are surprisingly similar when we're discussing these things. I'd say that in some other fields the differences are considerably bigger.
>
> (Export Manager)

After all, researchers have often worked abroad and use a lot of English in their daily work, which also facilitates communication with end customers. Furthermore, the scientific language unites researchers all over the world. Educating the end customers was thus a rather similar process, irrespective of the country in which they were located. However, the buying practices varied considerably: decisions were made by institutions such as universities in some countries, whereas in others they were made by individual researchers. Thus, a variety of approaches were used with customers in each country according to the knowledge that the local distributors possessed. In other words, the orientation of Finnzymes seems to have been mainly *polycentric* at that stage.

The close contacts with distributors have been appreciated, and communication with them has been regular. Bigger meetings tend to be organized in conjunction with trade fairs. The distributors are visited regularly and given information about the new products, and the visits often also extend to the end users. Meetings with the end users have been important in terms of learning more about their needs and also of assessing the market knowledge of the distributor:

> It's also an indication to us about how well the distributors understand customers' needs, what the feeling is when they're meeting customers. Usually how it goes is that for example I'm there in the meeting, then there's the representative of the distributor and then the customer. You can easily see how well it's all working, whether the distributor knows the customer and whether they can act in the right way.
>
> (Export Manager)

Now when the market is mature, customer knowledge regarding the enzymes has increased. At the same time, the competition has become tougher. The biotechnology research field is developing rapidly, and new needs for enzymes have emerged. In order to better respond to these needs, Finnzymes has *reacted* and developed new products, some of them based on DyNAzyme I: e.g. DyNAzyme II and DyNAzyme EXT. Given the usage, customers need a wide variety of polymerases, and some may find the old ones the most suitable. For this reason, Finnzymes has kept the original DyNAzyme in its product range, and it is still, in fact, in great demand. It is considered a good and appropriate enzyme for the purposes for which it was originally developed.

The market for polymerases is likely to face considerable upheaval in the future. There is a trend toward the utilization of PCR as a diagnostic tool, which at first seems to indicate a growing need for polymerases, but at the same time, the amount needed for a reaction decreases steeply, which leads to diminishing demand. Finnzymes regards these coming changes as challenges and is continuing to monitor the market carefully. It gets information through its distributors, and since it distributes the products of its competitors in Finland, it is also in closer touch with the competitors' movements. It also actively follows the discussions in journals and on the Internet. As a small firm, Finnzymes considers its influence on the coming changes to be quite small, but it is striving to anticipate them. New significant applications are likely to arise in the future due to the coming advances in studying the functions of genes.

Aiming for growth

As already mentioned in section 5.2.2, Finnzymes was a small firm at the beginning of the 1990s with a turnover of 1.4 million euros and ten employees. In 2005, the turnover was 9,372,700 euros, and the firm had 60 employees. The organization has grown steadily, which, according to the CEO, has guaranteed the maintenance of the original organizational culture:

> I don't believe in the kind of building of a company where you can have 100% growth in a year and half of the staff is new. You can't build a culture in a company like that and that, in turn, becomes something nobody's able to control.
>
> (CEO)

Indeed, the spirit of the early years seems to have prevailed to some extent, since the organization appears to have remained rather *organic*. Over the years, the information flow has improved and management has tried to focus it on those concerned. Consequently, the knowledge sharing has become more systematic and structured, while the communication seems to have remained open. For instance, the communication between marketing and development is considered good, and emerging issues are efficiently taken care of.

As the firm has grown, the decision-making has become more dispersed. Decisions have always been made rather quickly without regard to hierarchical barriers. Subordinates have been involved, and decisions have been redefined and changed later if changing circumstances make it necessary. However, the CEO expressed the hope that employees would be more proactive and initiate new ideas: "I've tried to emphasize that we [managers] need more input, such as 'this is what we've been thinking lately, shall we start developing this?' More of that, there definitely isn't as much of it now as I'd like" (CEO). Hence, the proactiveness of Finnzymes still seems to rely rather heavily on the few key persons who set up the firm and developed DyNAzyme. Management has tried to *foster customer-related proactiveness mainly through enabling this kind of behavior.* Organizational structures have been built to enhance open communication, and mobility of employees between various functions inside the firm has been encouraged but not forced. Most people responsible for marketing and exporting are former researchers: the move from the laboratory to the marketing department is considered rather easy, since the latest research knowledge is a prerequisite in marketing:

> You can talk to customers about science and you get to know new people all the time. In that respect marketing is fascinating. If you sell the products of several different clients then they train you for it. There's a lot of technical information involved.
>
> (Export Manager)

The early decision of Finnzymes to recruit accomplished scientists for marketing is considered an advantage, since they understand the customer needs and can discuss them on the same level:

> Also the majority [of salesmen] have worked in a laboratory for a while and then there are also a lot of doctors who have even written their PhDs on this, which means that this kind of person might meet a customer who is just starting on a PhD when they've finished one themselves. [...] In this industry the salesperson might have been doing the same job with the exact same equipment as his customer, and can therefore speak from experience and be very credible [...] We've tried to make sure we have so-called A-class or deeper knowledge as much as possible. We've always valued it very highly.
>
> (Export Manager)

Incentive payments have been made to reward successful work. However, it became apparent in the interviews that although monetary rewards are important, they are not the main motivator:

> Money as an incentive ... Money keeps you happy for exactly two months. Being motivated at work comes from completely different things, that you're genuinely interested in those things. Natural sciences are incredibly interesting, you discover something new ... I mean, of course money's important to every one of us, I can't deny that, but it just doesn't work as an incentive.
>
> (CEO)

This is such an intriguing and interesting field to work in. I'd say that people are quite dedicated, prepared to work really, really hard in busy times. I don't feel that people are forcing themselves to do this, they have a genuine interest in it and they're interested in developing themselves.

(Export Manager)

Since natural scientists seem to be motivated by the desire to work in the field of science and to invent something new, the motivation in Finnzymes seems to be connected to innovation development. Disciplinary action is rarely used, and the turnover of employees is very low. According to the CEO, the reputation of Finnzymes as an employer is reasonably good, and people seem to enjoy their work.

The enthusiasm about the innovation also seems to have increased customer-related proactiveness, and researchers tend to be very keen to find solutions to customers' problems. It was stressed in the interviews that it was characteristic of natural scientists to search for answers to problems through trial and error, since there was always more than one solution. When inventing something new, researchers should be able to experiment, and experiments tend to fail. In its failures, Finnzymes has relied on open communication and learning from mistakes. The CEO has encouraged people to play with their new ideas:

We're interested in money. In all projects of course we're interested in the commercial product. But we can also become interested in things solely based on the fact that they seem interesting. We don't necessarily think first about how we could make money out of them. It's more a kind of curiosity. I think you'll be successful in this field if you can play around with things. You need to be able to play around and try different things. You just can't always think that money is the most important.

(CEO)

The CEO also emphasized how important it was to realize how far the firm could go if it wanted to guarantee its survival. It was considered crucial for the management to set the limits on these experiments, thereby assuring the profitability of the business in the long run. Finnzymes' financial assets have always been scarce but adequate. Acting as a retailer of tools for molecular biology has given it stability and the financial resources to develop its own products. If there had been more resources, they would have been invested mainly in product development equipment. Thus, it seems that additional finance would not have directly increased proactiveness toward customers.

Over the years, Finnzymes seems to have benefited considerably from *external linkages*. It has cooperated extensively, both locally and internationally. For instance, Tekes provided funding at the very early stage, and NEB (New England Biolabs Inc., USA) provided know-how for the enzyme production and marketing. Furthermore, knowledge transfer took place with various international universities. Openness with the external partners seems to have been rather extensive. The behavior of NEB in this context has been exemplary: the external cooperation has worked well and is continuing with the same partners. Finnzymes has used horizontal, vertical and diagonal cooperation, the vertical cooperation (with NEB) and the horizontal cooperation (with other distributors) in particular having fostered proactiveness toward customers.

In 2006 and 2007, Finnzymes made considerable investments in order to secure its competitive advantage in the future. It aims to create a group of companies that can offer a complete solution for the PCR market, including reagents, instruments and consumables. The first steps in that direction were the establishment of a new company, Finnzymes Instruments Ltd, and the consequent acquisition of Bridge Bioscience Corporation, which was later renamed BioInnovations Ltd.

5.2.7 Customer-related proactiveness during the DyNAzyme development process

In sum, the time between DyNAzyme ideation and market success was 11 years. During this time, the polymerase was researched, purified, optimized and established. These key events demarcate the different stages of the innovation development process, as illustrated in Figure 5.6.

The customer-related proactiveness seemed to vary along the process of DyNAzyme development (see Figure 5.7). The market opportunity at the idea-generation stage was mainly *anticipated*. The founders of Finnzymes were actively following the emerging trends in biotechnology and believed that if they could manage to create the best thermostable polymerase, it would quickly create a market for itself. Since the new advancements in the field of gene technology were transforming the whole industry, it was anticipated that there would be a global market for this kind of polymerase: in terms of scope, this anticipation could be called *geocentric*.

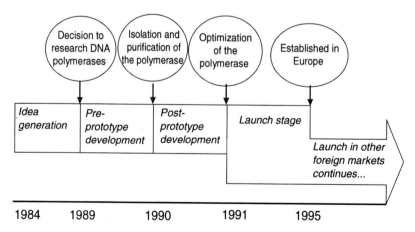

Figure 5.6 The innovation development stages of DyNAzyme.

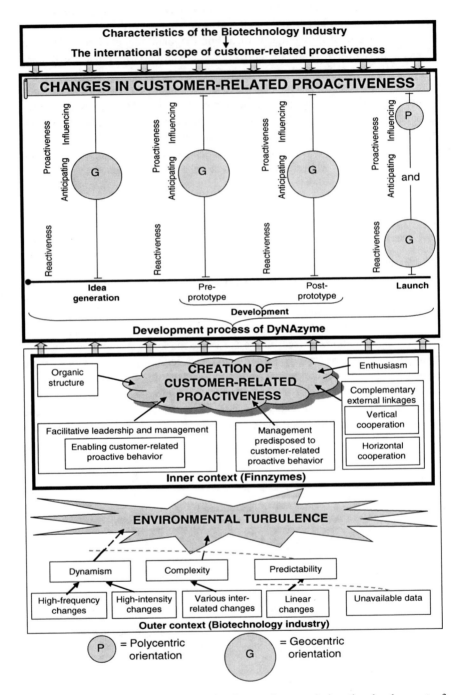

Figure 5.7 The change in customer-related proactiveness during the development of DyNAzyme.

This *geocentric anticipation* lasted throughout the development stage. Because of the nature of the product, it was not possible to gather feedback from the market until the launch stage. However, Finnzymes was able to rely on its own experience of restriction enzymes, and on its knowledge about the users, which it had gained when acting as a distributor for NEB's products. Since the needs of the researchers in the biotechnology field were considered global, the scope of the anticipation was geocentric. Because the *biotechnology industry and its characteristics seemed to influence the international scope* of customer-related proactiveness, it was further added to the framework.

At the launch stage, Finnzymes was actively *influencing* the market and creating a demand for the new polymerase. This was necessary since, by their nature, biotechnical trials increase the barriers to adopting a new polymerase: it was challenging to get the researchers to try DyNAzyme. Demand was created by influencing retailers and end users. Personal visits played an important role in awareness-raising and training in polymerase usage. Demand was also influenced through advertisements, scientific conferences and biotech trade fairs. Although the needs were rather similar across the globe, it was necessary to approach each country individually. *A polycentric* approach was justified by the need to find local retailers and by the distinctive buying practices in different countries. However, at the launch stage, Finnzymes was also *reacting* to customers' requirements: the biotechnology field was developing quickly, and new needs for polymerases were emerging. Finnzymes reacted to these by producing derivatives of DyNAzyme (and also other new products). Since the requirements were considered global, the scope of the reaction could be called *geocentric*.

The environment was not particularly *turbulent* during the development of DyNAzyme. The biotechnology industry was experiencing various big, fast-occurring, interrelated changes, which could have increased the turbulence, but this was mitigated by the ability of the firms operating in the field to anticipate in which direction the changes were going and what kind of innovations would be needed in the future. Hence, some changes were linear.

The inner context of Finnzymes fostered customer-related proactiveness through its organic organizational structure and by enabling customer-related-behavior in the organization. Furthermore, both vertical and horizontal cooperation were utilized to enhance proactiveness toward customers. The managers were apparently proactive personalities, and although primarily oriented toward the innovation, their enthusiasm seems to have made them behave proactively toward customers.

Unlike in the theoretical framework, building up customer-related proactive capabilities and motivating this kind of behavior was not specifically emphasized in Finnzymes. Furthermore, the organizational culture and liquid financial assets did not particularly enhance proactiveness toward customers. Apparently, all these elements were utilized to increase product development capabilities and scientific research. Norms, values and artifacts prevalent in the firm were more related to scientific issues and research work than to customer needs. This emphasis on innovation development instead of on customer requirements was probably necessary in order to stay ahead of developments in the high-tech field, in which novel technologies and products are continuously revolutionizing current practices and market structures.

The case demonstrated the fact that, even though the product-related issues were more emphasized in Finnzymes, the *enthusiasm* that started with the product development also increased proactiveness toward customers. It seems that even if the firm decides to invest its scarce resources in product development, it may still be able to foster customer-related proactiveness provided that the enthusiasm related to the innovation spreads from its developers throughout the whole organization. Enthusiasm was thus added to the original framework.

5.3 ePost Letter

5.3.1 ePost Letter as a radical innovation

Written targeted communication has a long history: camel post was introduced in Egypt way back in 280 BC. Letter writing increased considerably along with the establishment of public postal services. The Finnish postal service was founded in the early seventeenth century, and the volume of letter mail, particularly business letters,[10] grew rather steadily until the early 1990s. However, in recent decades, traditional letter mail has been increasingly challenged by a number of new modes of targeted communication, including the telegram, telex, telefax and e-mail.

During the late 1980s and early 1990s, Finland Post developed an automated letter-mail-handling system and service called ePost Letter, a novel combination of electronic communication and traditional letter mail: the sender transfers the data in electronic form to the national ePost sorting center. The data delivered to the Post consists of the letter content (including the address) and control codes for the desired services in its automatic processing. The data is sorted and merged in the sorting center and distributed in electronic form to the sorting center nearest to the recipient. There the letters are printed, enveloped and injected into the local postal-delivery system. Hence, ePost is delivered to the recipient as ordinary letter mail (see Figure 5.8).

It is worth emphasizing that, in this study, the term "ePost Letter" refers only to the automated letter-mailing system and service, since it fulfills the criteria of a radical innovation. However, in other connections, it may include conventional enveloping and printing services that Finland Post offers to its customers. Since the development and marketing of these services played a pivotal role in ePost Letter development, some aspects of them are described in the following sections.

The main emphasis shifted from ePost Letter development to marketing in 1992, and ePost Letter turned out to be a *commercial success*. The volume has been increasing steadily (see Figure 5.9), and more than 80 million were sent in Finland in 2006. Furthermore, the innovation also brought substantial cost benefits to Finland Post, since printing in the local ePost centre diminished the need

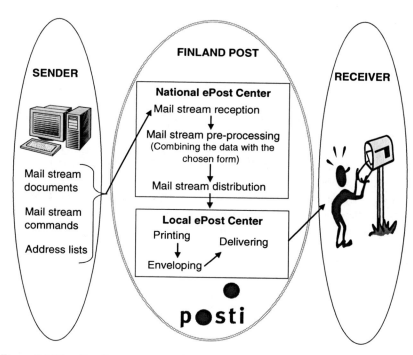

Figure 5.8 The ePost Letter system and service.

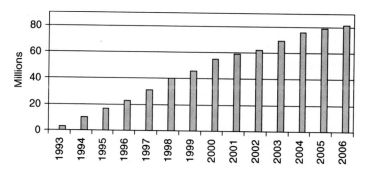

Figure 5.9 The volume of ePost Letters in Finland 1993–2006 (Finland Post Annual Reports 1993–2006).

for long-haul road transport and enabled the printed letters to be sorted by postal code. In addition to a new source of revenue and resource saving, the ePost Letter also brought image-related benefits to Finland Post. "For image reasons ... that is, to show that the Post isn't an old-fashioned and stuffy institution still stuck in the era when people used quills to write, but that we're on the cutting edge of development" (Product Manager, Atkos).

With hindsight I'd say that it gave more self-confidence to Finland Post and a feeling of "we can do this too!" [...] Our position is a lot better at the moment, compared with what it would have been without this. It has been surprisingly significant strategically. In a way it has been a clear confidence booster for Finland Post.

(Director, Letter and Direct Marketing Services)

Hence, ePost Letter gave the old postal industry a new channel through which to compete in the Internet age, and it was called the "Flagship of Finland Post" in the 1990s.

The technological novelty in ePost Letter stemmed from the system, which combines ePost software and the printing, enveloping and delivery system. The customer can include in one single mail stream a large variety of different kinds of letters, and the system can treat, combine, print and distribute various mail streams in a way that utilizes substantial economies of scale. The technological novelty was also recognized in 1993 when the international electronic documentation systems organization Xplor International awarded it "The Innovator of the Year" prize.

ePost Letter provided *substantial benefits* for the buyers of the service, since it was likely to save time, trouble and money. Time-saving resulted in speeding up invoicing and releasing resources from posting routines (e.g. printing and enveloping) to more productive work. Along with the decrease in paperwork, the need to purchase and store invoice forms and envelopes and to file paper invoices disappeared. Thus, ePost worked to rationalize the customer's operations and to save storage, filing and employment costs. It also enabled customers to conduct high-volume mailing at lower prices and with shorter delivery times and made economies of scale in posting accessible to smaller firms. Consequently, by abolishing certain posting routines, ePost Letter *changed the behavior* in the adopting firms.

ePost Letter has other characteristics apart from innovativeness that influence its development and marketing. First of all, it is *a service*. It consists of various activities, it is produced and consumed simultaneously, and the customer participates in the service production. Consequently, as far as the customer is concerned, it is mostly intangible and difficult to evaluate. More specifically, it is *a high-tech continuously rendered service*. Hence, it is based on the use of information technology and involves a continuous flow of interactions between the customer and the Post (cf. Grönroos 2000: 47–50; Miles 1994: 244–245). Furthermore, the ePost Letter service is bought by other *businesses* in order to externalize activities that have been partly conducted inside the organization, and it is often utilized in invoicing. Thus, it handles *confidential information* that is directly related to the *revenue generation* of the customer firm, which makes it an important service.

It was not only a service that ePost Letter provided, it also created a new *system* through which the service could be produced. This system can be sold to other national postal operators. Finland Post has strived to benefit financially both from the service and from the system selling.

5.3.2 *Finland Post*[11] *– the developer of the ePost Letter*

The Finnish postal services started in 1638. They were initially part of the Swedish Royal Mail and later on operated by a central office in the Grand Duchy ruled by the Russian Tsar. Along with the independence of Finland in 1917, the Finnish postal system became one of the symbols of the new nation. As a state-owned organization, its role in developing the infrastructure of Finnish society has been regulated by law. Its operations include postal services and telecommunications.

Over the years, the structure of the organization has gone through substantial changes. It has been slowly establishing its autonomy since the 1970s, and efficiency, productivity and new technology have been accentuated as it has strived to acquire a more modern image and to become more market-oriented. It was still a government department in the 1980s and closely constrained by the national budget. Thus, its income (i.e. postal charges) was fixed and its expenditure was predetermined in detail, which influenced the organizational behavior. The budget was closely monitored, and it was difficult to recruit new employees, for instance.

In 1990, the Post became a more independent establishment, a public corporation, and its operating expenditure had to be covered by its own income. However, the Government still set the prices of its basic services. The organization was reformed in line with the new business strategy. In 1994, just when Finland fell into economic recession, the Post became a limited-liability company, PT Finland Group, which was composed of both postal and telecommunications services.

In order to follow the European examples and to be able to privatize its telecommunications services, the PT Finland Group was divided into two companies, Finland Post Ltd and Telecom Finland Ltd, in 1997. Telecom (from 1998 Sonera Ltd, later TeliaSonera) was partly privatized, and Finland Post remained state-owned, although an independent limited-liability company. Thus, the Finland Post Group has gone through *a succession of big organizational changes* in recent decades.

The postal service has been one of Finland's biggest employers, although the number of employees has decreased in the last decade: the Finland Post Group had 24,806 employees and a turnover of 1,551 million euros in 2006 (see Figure 5.10). Its large size and state-ownership have fostered formally laid-down procedures and hierarchical decision-making, and flexibility has thus been at a rather low level. As one interviewee pointed out, it has been "a big ship that is not quick to turn around."

Thus, as an organization, the Finland Post seems generally to have been rather conservative and mechanistic, although inside it there were other ways of doing things. ePost Letter was largely the brainchild of Jarmo Koivunen, who could be described as a long-term postal worker. He started to work full time at the Post permanently in 1966, but even before that he had occasionally been employed as a summer trainee. He started the ideation that led to the ePostLetter

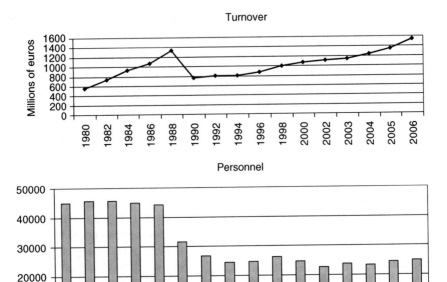

Figure 5.10 Turnover and personnel of Finland Post. (Note that in 1990 the Post was converted from a government department into a public corporation.) (Posts and telecommunications of Finland 1980–1988; Finland Post Annual Reports 1990–2006).

in 1982. During the development stage, he was the director of the Postal Electronic Mail and Mailing Services (PEM) unit and later became managing director of IDP (International Data Post). Koivunen's determination and personality seem to have played a pivotal role in the creation of ePost Letter. A former working companion described his personality as follows:

> Product developer. Engineer. A great personality with a good sense of humor. And an iron will, I've never met such a stubborn person. But I have to say that there's no one else than Jarmo who could have made it in the circumstances at the time. He really had such an iron will that he would have walked through stone walls if necessary
>
> (Customer relationship director)

As a Master of Sciences in engineering physics, Koivunen was interested in computers and had worked with them since the 1960s. In the Post, he was able to use them and do work that inspired him, which made him very enthusiastic about his work. As he said himself:

> All my working life I've been lucky to be in roles where I've been very motivated and enthusiastic – it was so hard to go home on Fridays when I had to wait until Monday to continue, because we didn't have the equipment for working from home at the time. Of course, you could take documents home and think about them. [...] They were such fun times back then. That's why I liked the job, because I was almost making the decisions at the same time as well. I didn't really have the kind of managers at the time who would have resisted it. They were just happy when someone suggested something.
>
> (Former director of the PEM unit; Former managing director of the IDP)

Thus, the director of the PEM unit seems to have had a *proactive personality* and also the characteristics of customer-related proactiveness: although he was extremely busy in making the ePost Letter system technically functional, he regarded contacts with customers as both beneficial and fun:

> I'm really enthusiastic, because I like to talk a lot, and I like people and cultures and the like, so ... When they had the office technology fair, I was there as well. I just stood there next to the aisle trying to lure people in and I thought it was great fun! I've always liked to market and sell things to people.
>
> (Former director of the PEM unit; Former managing director of the IDP)

As he was developing ePost Letter, Koivunen began to gather other people together in his team, which then became the Postal Electronic Mail and Mailing Services unit (later referred to as the PEM unit). He was regarded as supportive. He told his people what he wanted them to accomplish, but he did not give detailed instructions on how to do it. Thus, the emphasis was on getting things done. Teamwork was encouraged and the employees were willing to bring up new ideas. People were *enthusiastic* about the development of ePost Letter, and it seems that the enthusiasm spread from the PEM unit to the sales force, who conveyed it further to the customers:

> Jarmo had such an inspiring style. And when I went through the whole thing and realized what an opportunity this was, I could hardly hold my excitement. There was even one customer who said that he couldn't help getting excited when he saw that someone had a spark like that. That spark then spread throughout the organization. That was one of the main things that when you get people onboard, and there are a lot of them, then it can't fail. I think everyone who was selling the ePost Letter at the time had that spark.
>
> (Customer relationship director)

The organization was considered very informal and open in its knowledge sharing:

> I had all that support, which was very important as well. To know that you're trusted. That you've got all these targets and then you've got the freedom to go and do it. That you've got the freedom and that you can think for yourself. I probably had exceptionally good conditions at the time, I think.
>
> (Customer relationship director)

It seems that the PEM was much more *organic* than the Post in general. Furthermore, the director apparently *enabled and motivated proactive behavior toward the customers*. Working in a unit that was regarded as highly international, and as a pioneer in the utilization of new technology, motivated the employees to work there and also attracted new employees.

> Really, the ePost Letter itself and Jarmo's unit in general were the so-called flagship for Post in a sort of a technological sense. It was great to get to work there, simply working there made you motivated. And Jarmo was internationally inclined. There were international meetings ... it was all very international. This was a characteristic that attracted what you might call Post's best resources to this unit. It was really engaging.
>
> (Director, Letter and Direct Marketing Services)

Employees who were extrovert and both customer- and technologically oriented liked to work in the PEM unit. The team consisted of people who were driven by the challenges and motivated to develop something new: "There was a fairly large group, the kind of people who you'd call movers and shakers. We perhaps took some risks, and just went for it and did it, so that we could make ePost Letter go forward" (Customer relationship director).Thus, it seems that it was not only the director but also the others in the unit who were *predisposed to customer-related proactive behavior*. Since the unit was small (eight people at most), it was important to have employees who were able and willing to do all kinds of work, and there was no room for hierarchical and formal behavior. Those who did not adjust to this way of working quickly left. Since there were instances of unsuccessful recruitment, it seems that proactiveness was not emphasized in the selection of new employees. In general, the Post invested considerably in training its employees, even during the 1990s when the Finnish economy experienced a deep recession. The training included courses on negotiation, presentation skills, time management, project management and selling skills, but it did not cover proactive capabilities. Thus, the *generation of proactive capabilities was not particularly emphasized in the organization*.

From the 1980s the importance of customer-orientation in general was recognized and promoted throughout the Post organization. Thus, the organizational culture was slowly changing from government service toward customer service, and the new orientation was emphasized in its values and norms. The PEM unit was no exception, although it seems that when the Post in general was emphasizing the fulfilling of stated customer requirements, the unit was concentrating more on anticipating and fulfilling future needs. This kind of thinking was reflected in its values and norms and was also apparent in certain artifacts, i.e. brochures emphasizing the benefits of the service and stories about how the service was sold to certain individual customers. Hence, it seems that the organizational culture supported customer-orientation in general, and the culture of the PEM unit in particular fostered *proactiveness*.

People in the PEM unit tended to be motivated mainly by the ePost Letter idea itself, whereas in the selling unit, the incentives included financial bonuses and selling competitions in which the best seller was rewarded with a trip to Paris or Rome, for example. These were regarded as strong incentives.

However, the freedom of action and resources granted to the PEM unit were largely dependent on the general Postal organization, and especially on the senior management. Management attitudes toward the development of ePost Letter have been both negative and positive, changing according to the interests of the respective CEOs. Thus, obstacles and hindrances have prevailed:

> There was some resistance to change, the sort of old-fashioned preference for the way things used to be ... And belittling ... Very prejudiced about the fact that nothing should ever change. The Post should stay absolutely the same. Everything should be the way it used to be. We'll do it the way we've been doing it for a hundred years. It was like everything that represented change was experienced as threatening.
>
> (Customer relationship director)

> Of course we were really enthusiastic about our jobs and that also showed. It was somewhat frowned upon, people seemed to think that we were looking down our noses at them and that we were talking about things they didn't understand. Some also thought that this wasn't really something that should concern us as a company. That it belonged to the IT department, or Tele or somewhere else, but definitely not to Post. That Post shouldn't have anything to do with things like these.
>
> (Former director of the PEM unit; Former managing director of the IDP)

Although there were aspirations inside the Post to concentrate all computer-based services under Telecommunications, senior management supported the viability of the PEM unit, since it considered the development of the ePost Letter system to be important. Of the CEOs, it was Pekka Tarjanne (1977–1989) and Pekka Vennamo (1989–1998) who were particularly enthusiastic about it. However, on the operating level, some workers saw it as a threat to their jobs, and they had to be convinced that it was actually generating more income for the Post than ordinary letter mail.

It was like "hey, we're selling much more now!" We're getting all of the postage revenue, and also the revenue from the material and from all these extra services we could offer to the customers. It was like we had to sell the idea internally, as well.

(Customer relationship director)

Even though resistance to change and individual fear of the unknown were slowing down the development of ePost Letter to some extent, there were also those who encouraged and supported it, particularly the sales people who received positive feedback from the customers. Support was important, since there was continuous competition between units and departments for investment allocations. The development of ePost Letter was justified in the investment negotiations, through the need to simultaneously increase customer-orientation and cost-efficiency in Finland Post, and encouraged by the new challenges and opportunities that arose in the firm's environment.

5.3.3 Environmental turbulence

The postal industry encountered considerable changes in the last two decades of the twentieth century. Traditionally, national postal operators have been the main players in the industry, leaving only a tightly controlled and limited arena for commercial transportation and delivery enterprises. The services have been strictly regulated, since their role in community social-welfare systems has been generally acknowledged and emphasized. Accessibility to postal services is guaranteed to all parts of the country under Finnish law. The Ministry of Transport and Communication (currently through the Post and Media Department) ensures compliance with the Law on Postal Services and with other regulations issued under its provisions.

Furthermore, global postal services have been influenced by international organizations. The Universal Postal Union (UPU), a United Nations specialized agency that sets the rules and recommendations for international mail exchange in order to ensure a universal network of postal services, has played a prominent role on the global level. On the regional level, the arena for dealing with postal and telecommunications issues was the European Conference of Postal and Telecommunications Administrations (CEPT), until it became an intergovernmental organization in the early 1990s. Consequently, European public postal operators founded PostEurop in 1993 to strengthen their cooperation. Finland Post is also a member of the Nordic postal group, NordPost, and it has been an active member of the Paris Group, which has aimed to increase the technical capabilities of postal organizations. It has also been involved in various standardization committees and development workshops. Hence, it seems that international cooperation among postal organizations has been multifaceted, and Finland Post has been an active participant. In particular, cooperation between the Nordic countries has been very close.

Although national postal services were operating in a regulated and protected environment at the beginning of the 1980s, there were major changes on the horizon. There was increasing pressure *to privatize* the national services and *open up the market for competition.* Traditionally, Finland Post had the sole right to deliver letter mail for a charge throughout the country, but the market was deregulated in 1990, and the right to an independent pricing policy was conferred in 1992. All of its operational activities were opened to competition in 1994, as long as uniform and corresponding services were guaranteed at a satisfactory level to all citizens.

The growing competition in the 1990s changed the nature of the relationships between the National Posts: former cooperators became potential competitors. The integration of the EU and the consequent EU efforts at postal deregulation further increased the competition. Indeed, it was predicted as early as in the 1980s that, sooner or later, new competitors, both domestic and international, would emerge. Given the diversification strategy (cf. Ansoff 1958), it also became increasingly evident that Finland Post should seek *growth through new services and new markets.*

Although the demand for communications, logistics and delivery services was increasing, the long-term outlook for traditional postal services was uncertain. It was already clear in the 1980s that *electronic communication* (e.g. telefax) was reducing the need for letter services, and when e-mail technology appeared in 1990, it gained a strong foothold in data transmission at the cost of conventional letter mail. On the other hand, the new technology also fostered the demand for postal services, since computer databases and growing direct marketing increased the amount of business. The projected growth of computer-based communications was seen in Finland Post as both a threat and an opportunity for traditional postal services:

In a way we also realized that electronic communication was on the rise. Telefax was already increasing in popularity and we thought that if people have landlines, they'll soon have faxes at home and everything will go through phone lines. We thought there was a demand and even pressure to maintain the competitiveness of paper communication. But on the other hand, we realized that technology evolves and that we, too, have to take advantage of it.

(Director, Letter and Direct Marketing Services)

In conclusion, the environment of the postal industry seems to have encountered significant but slow (i.e. high-intensity, low-frequency) changes in the 1980s and 1990s. Hence, *the dynamism* was somewhat high. However, the changes seem to have been interrelated (e.g. competition increased on account of the liberalization and privatization), which increased *the complexity* (see Figure 5.11).

Nevertheless, the above-mentioned changes have been linear, and there were plenty of data available on them beforehand, thus the predictability presumably decreased the environmental turbulence. They did not come as a surprise to

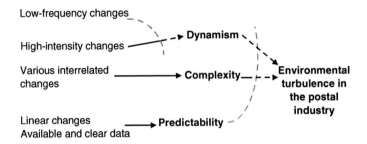

Low-frequency changes

High-intensity changes ⟶ Dynamism

Various interrelated changes ⟶ Complexity

Environmental turbulence in the postal industry

Linear changes
Available and clear data ⟶ Predictability

Figure 5.11 Factors contributing to the environmental turbulence in the postal industry (the degree of unity of the arrows illustrates the degree of influence of the factors).

Finland Post, at least, and the ideation of ePost Letter appeared to have been one of the ways in which it was preparing itself for the coming changes.

5.3.4 Idea generation

The beginning of the 1980s saw the emergence of various prognoses suggesting that the volume of conventional letter mail would not grow in the future. In particular, it was predicted that firms would send an increasing amount of correspondence by telefax and e-mail or through Electronic Data Interchange systems. This was a particularly severe threat to Finland Post, since letter mail accounted for around 48 percent of the postal-traffic revenue in the early 1980s. The organization made a determined effort to search for opportunities to benefit from the new technology and saw considerable potential in hybrid mail, i.e. mail in which the input in one communication medium is converted for delivery into another.

There were basically three lines of opportunities in that field: postfax, videotex and electronic mail. Postfax was based on the idea that those who do not own a telefax can go to the post office and send their faxes from there. They could be sent directly to the fax device of the recipient or, if he or she did not have one, they could be sent to the nearest post office and either delivered or picked up. Videotex is based on the idea that television screens installed in post offices could be used for communication. Electronic mail, i.e. mail sent in electronic format, was considered a relevant option for postal transmission in the more distant future.

When Koivunen started his work on developing hybrid mail services in December 1982, his predecessor urged him to devote 90 percent of his time to postfax, since at that time it was considered much more important than the other new media. However, Koivunen felt that the market for postfax services would decrease, since firms were increasingly buying their own fax machines and since the price of the service was too high for private persons. Furthermore, he calculated that the service would actually require too many fax machines and too many employees in each post office, which would not make it a viable solution in practice. He therefore concentrated his efforts on finding ways of combining electronic and conventional mail. Even though he assumed that electronic mail would be the standard medium for sending letters in the future, he saw hybrid mail as a viable and profitable solution for the transition stage.

At the same time, he was asked to find a new enveloping machine. Finland Post started to offer a small-scale mailing service in 1973; in other words, it took care of enveloping unaddressed mail on behalf of certain firms. At the beginning of the 1980s, Shell asked them to take care of the enveloping of their invoices as well, and thus there was a need to find an enveloping machine that could also handle addressed mail. When he found a suitable machine, Koivunen realized that the next phase would probably be to provide printing services for customers. As a result, the printing service began in 1985.

The printing-service customers were large firms with big consumer markets and direct invoicing (e.g. electricity plants and oil companies). Given their letter volumes, the utilization of the printing service was cost-effective. The data to be printed was delivered in magnetic files, but the idea of receiving it in electronic form directly from the customer's computer system was also becoming increasingly attractive:

> What good is it to build a four-lane motorway and suddenly there's a set of traffic lights and bollards and the motorway changes into a cart track? That's the situation we're in now. Customers have computers, fancy systems that produce massive amounts of letters. The Post then receives them in its own old-fashioned, traditional way in mailboxes. Then the letters are poured into some big buckets from where the employees sort them, turn them the right way around, and...
> (Former director of the PEM unit; Former managing director of the IDP)

Thus, it was anticipated that there was a latent need for this kind of ePost Letter service. The customers wanted an easier way to mail their letters, and it was assumed that this would enable the Post to take care of some mailing-related tasks for the customer.

> It was more of a hunch, really. Of course we'd considered it quite carefully, in terms of the demand for letters. We had estimates of how electronic communication would start replacing letters at some point. In a way, we already realized how easy it would be in terms of billing, as in the ease of transferring digitized information. We figured that if, say, a third of the letters are bills, and a lot of them are already in a digital format. Bank statements were like this as well. That's pretty much the market for this kind of service, really and that's the general idea we had.
> (Director, Letter and Direct Marketing Services)

Furthermore, a system that could receive data directly from the customers' computers also seemed to be suitable for smaller firms that were doing their invoicing manually. It was assumed that if the system could combine various smaller letter streams, it would create economies of scale for both the sender and the Post. In sum, customer-related proactiveness seems to have been rather high at the idea-generation stage, since the *opportunities in the market were anticipated*. Given the nature of the postal industry, the picture of potential ePost Letter Service customers was *ethnocentric*, i.e. restricted to the Finnish market.

On the international arena, the new ideas were discussed among the Paris Group members. The Paris Group comprised representatives of around 25 national postal-service organizations with the most advanced systems. They met twice a year and pondered on the future developments of postal technology. The director of the PEM unit attended the Group meetings, which he describes as follows:

> These meetings with the Paris group ... there was a lot of talk! They had a principle of nobody marketing or selling anything to anyone – instead everyone speaks about what they've been doing and others may pick up good ideas if they wish.
> (Former director of the PEM unit; Former managing director of the IDP)

The meetings gave him an opportunity to form close contacts with foreign postal service organizations, especially in the Nordic Countries. However, he was among the few who believed in the opportunities of hybrid mail. Most of the members were interested in modern technology, especially in postfax and text-based messaging, thus his ideas of combining traditional mail with e-mail did not arouse much enthusiasm. He nevertheless assumed that there might be an international market for the ePost Letter system, although he acknowledged that its heterogeneity could be an obstacle to its global diffusion:

> We also went to Turkey and asked them "how many postal codes do you have – you do use postal codes, don't you?" "Yes", they said, "of course we do." They also said they'd done some research and found out that only about 6% of the letters had postal codes on them, and that they were almost always incorrect. Considering the fact that Turkey is not exactly insignificant in terms of the extent of its postal organization, the group of countries that could be potential buyers was pretty small.
> (Former director of the PEM unit; Former managing director of the IDP)

Thus, the international opportunities for the ePost Letter system were considered to be limited to certain countries with the most advanced postal systems, i.e. the international scope of customer-related proactiveness seems to have been *polycentric*.

5.3.5 Development

The big challenge was in getting the separate functions, i.e. data input from the customer, data processing, printing, enveloping and delivering, to work as a system. The target was to create a system to provide a new kind of service that would be "irresistible to customers, inaccessible to competitors and profitable to the Post." The development work began in 1986 when Koivunen and a few programmers devised the programming for the first ADP system in isolation from the Post's information-technology unit:

> Jarmo didn't really trust them, or maybe he just didn't get a lot of support from there in the beginning. Then he just decided to develop it himself, which was pretty much the start of a sort of a "home bakery" model, where Jarmo hired his own IT experts whereas the IT unit had their own ... the two never really discussed things. It was perhaps the biggest gap in Post's organization. The so-called Post IT department and Jarmo's IT department were separate ... The distrust was there from the beginning. Of course there were matters of scale, and Jarmo realized that it just wasn't possible to develop the project in that environment. I'm sure he was absolutely right. And then on the other hand, here in Post's big IT department they thought that synergies weren't being made use of. This is what Jarmo constantly had to battle with, that they were able to maintain their independence and that way maintain the flexibility of the development. It just wouldn't have been possible within Post's general framework.
>
> (Director, Letter and Direct Marketing Services)

Thus, Koivunen's own team produced the first prototype of the system in 1987, created in Pascal programming language with a Microvax computer. The emphasis during its development was on the technological features, and the needs of customers played a minor role. It was considered important to get the technology working first, and besides, the developers assumed that the benefits of ePost Letter would be so considerable that the service would sell itself. This assumption was supported by the Post's previous experience in providing a printing service for customers, when even detailed requests had become routine. The fact that this experience was before the prototype development made it possible to *anticipate* the needs of Finnish customers for an ePost Letter service.

There were also customers to be found abroad, and although the system was developed for Finland Post to serve Finnish customers, even at the development stage account was taken of the possibility of selling the system to other National Posts. Thus, the developers also tried to anticipate the needs of these international customers. However, as stated above, there was a limited number of countries with sufficiently well-developed postal systems, and thus, the international scope was *polycentric*.

Although the prototype was developed inside the firm, the ADP programming was subsequently bought from IBM.

> In reality we wouldn't have got very far with our own prototype. We were able to use it to prove to ourselves and the decision-makers that this really works and this is what the concept is like, and we could also tell clients that

this is what we've come up with, but IBM is going to make us the real, commercial product.

(Former director of the PEM unit; Former managing director of the IDP)

The cooperation with IBM was important, since they had created what was at that time the only printing architecture that could also be used to print pictures (i.e. Advanced Function Printing). Hence, only they knew the architecture thoroughly, and they also lent credibility to the system, especially on the international market: "Well, if we think about this kind of commercial system that you start selling to others ... Nobody's going to buy it if we say we've built it ourselves. But if we say that IBM's done it, then..." (Former director of the PEM unit; Former managing director of the IDP).The cooperation with IBM worked well. The effectiveness of the system improved during the development process, and the first IBM version was ready in 1990.

As things progressed, it became more and more important to know what kind of requirements the customers would have for the ePostLetter service. Since the beginning of the 1990s, Finland Post has been conducting regular market research on the letter business in terms of who is sending letters, who is receiving them and what kind of data they contain. Furthermore, the sales force was of significant help in giving information about customers and their needs, which was crucial in the further development of the system. Direct contact with customers facilitates *reaction* to their requirements:

Customers did a lot of the development for us. To be honest, we had enough on our hands in breaking new clients in technologically, so to speak ... And then we had their ideas and we had to go through them as well, process them and find the good bits and add them to the existing system.

(Former director of the PEM unit; Former managing director of the IDP)

For chaotic product development, that's very typical ... that we went through several systems, or technical solutions. We constantly had to go through ideas, demands and hopes pouring in from everywhere. We were in a constant client dialogue, and that's how they turn up. The ideas also come from inside the company, from sales for example, and in a way we tried to accommodate them then.

(Development manager)

Thus, after the prototype development, the needs of the end customers for the service could be *reacted to*. The customers at that stage were major letter senders, such as Shell and Nestlé. They sent over 5000 letters/batch, and mainly utilized the enveloping and printing services (see Figure 5.12). It was important for the PEM unit to attract big customers so that it could increase its earnings and justify the further development of the ePost Letter service. These users delivered the data mostly in magnetic files or on diskettes. Delivery through modems was slowly gaining ground, although they did not work particularly

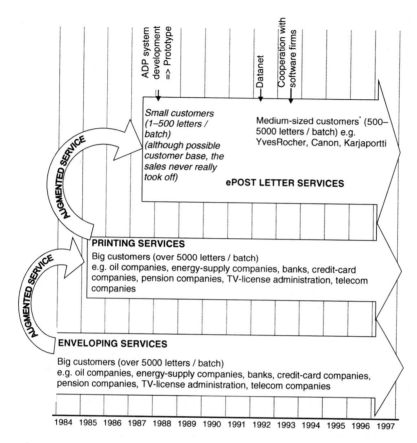

ADP system development => Prototype

Datanet

Cooperation with software firms

AUGMENTED SERVICE

Small customers (1–500 letters / batch) (although possible customer base, the sales never really took off)

Medium-sized customers* (500–5000 letters / batch) e.g. YvesRocher, Canon, Karjaportti

ePOST LETTER SERVICES

PRINTING SERVICES

Big customers (over 5000 letters / batch) e.g. oil companies, energy-supply companies, banks, credit-card companies, pension companies, TV-license administration, telecom companies

ENVELOPING SERVICES

Big customers (over 5000 letters / batch) e.g. oil companies, energy-supply companies, banks, credit-card companies, pension companies, TV-license administration, telecom companies

1984 1985 1986 1987 1988 1989 1990 1991 1992 1993 1994 1995 1996 1997

* Recently, from the turn of the millenium, big customers (e.g. banks and public administration) have also increasingly begun to use the ePost Letter service.

Figure 5.12 The introduction of special features of the enveloping, printing and ePost Letter service and the related customer base in Finland.

well with big batches. Nevertheless, it was thought that modems would facilitate ePost Letter sending in small firms, who could send the data directly from their PCs.

Organizationally, the PEM unit was in rather a challenging position during the prototype development: it was developing a new ePost Letter system, but it had no direct control over the ePost Centre, i.e. those who were carrying through the production tasks of the system.

I had no authority there whatsoever, everything went through official routes. And that meant that I had to have good relations with the managers there, as I worked closely with them, and it worked well. There were some really good guys there, although not educated enough, and they didn't understand the technology sufficiently. But then again, it was more important at the time that we had the support of the people who were actually doing it, who were on the front line. To have them behind the idea. And they were really enthusiastic about it, there were constantly people wanting to work in that unit.

(Former director of the PEM unit; Former managing director of the IDP)

Hence, the relationships between the PEM unit and production were considered good, although there were naturally occasional slight disagreements over practical issues. Although the ePost Letter System took care of the sorting and thus alleviated the workload of production, the PEM unit did not seek compensation, since they did not want to put a strain on relations. When the prototype was ready, control over the production was given to the PEM unit. The printing and enveloping services generated enough financial resources to enable ePost Letter service development, and besides, the senior management supported money allocation to this project, which they considered very important. Consequently, the financial resources were considered sufficient for the ePost Letter development, especially since the advantages of proceeding at a pace that allowed the developers to retain control were acknowledged:

It wasn't about money, but about … in the end, it was a really big and new idea, which meant that we had to leave it alone slightly and let it progress at its own speed. What I considered important was that we hold a tight rein over it at all times, that we don't lose control of what's happening. I knew it would go all wrong if that ever happened. And the more people we would've hired, the more difficult it would've been to work on this – it was a lot better that we had a good, talented small team that knew what it was doing and everyone knew how to work independently in the same direction.

(Former director of the PEM unit; Former managing director of the IDP)

The development team seems to have enjoyed their work. They were doing something that was completely new, and the spirit was considered good.

Once the prototype had appeared, the Nordic countries started to pay more attention to the system, and in 1990, representatives of the Nordic postal organization started to think about arousing international interest in ePost Letter. They decided to organize an international ePost Mail[12] Conference, which was done independently of the Paris group, whose members had not generally expressed interest in ePost Letter. The conference was held in August 1991 in Finland. Invitations were sent to approximately 180 national postal operators around the world, even though the main focus was on representatives of those in Germany and France, who were considered to be the most likely potential customers.

Although most of the participants were Finnish, the conference proved to be influential in forming international contacts and raising awareness about the new system. It was the first strong attempt to deliberately *influence* certain national postal operators, thus exhibiting *polycentric* proactiveness. The influencing extended to unofficial occasions that took place in conjunction with the Paris Group meetings, for example. These unofficial meetings were considered especially important in creating trust between the business contacts.

In sum, the needs of the customers of the *service* seem to have been mainly anticipated before the prototype development stage. It was after this that it became possible to react to the customers' needs and to involve them in the development work. Given the nature of the postal industry, these customers were Finnish, and thus the scope was ethnocentric at this stage. However, as far as customers of the *system* were concerned, i.e. other national postal services, the scope was polycentric and directed toward certain countries with advanced postal systems. Their needs seem to have been mainly anticipated before the prototype development, but then Finland Post started to influence them and actively to create awareness of the system.

5.3.6 Creating the market

The emphasis shifted from the ePost Letter system and service development to marketing in 1992. At that time, Finland Post was establishing the Finnish infrastructure for the system and also starting the intensive launch of the service in Finland. Finland was then in a deep economic depression, which enhanced the externalization of corporate activities and created more pressure to minimize the turnover of receivables. Consequently, firms strived to rationalize their mailing costs, which could be seen in the considerable change from first- to second-class mail. These attempts to decrease costs increased interest in the ePost Letter service. Thus, the timing of the launch was also particularly fortuitous.

The international launch of the ePost Letter System in 1992 began through a newly established firm, Nordic Data Post (NDP). NDP bought the system and took care of developing it further, thus enabling Finland Post to concentrate on marketing the service to Finnish customers.

Creating the infrastructure for the ePost Letter System in Finland

In order to take full advantage of the benefits of the ePost Letter System, local ePost Letter centers (Helsinki, Turku, Tampere and Oulu) were connected together via a data-communications network.[13] The system was almost in place at the beginning of the 1990s, and there were already some customers for it – large firms that had utilized the enveloping and printing services. However, in order to find new ones, certain technical boundaries had to be crossed. Customers wishing to use the ePost Letter service had to send the data in a certain fixed-access format. This was not usually a problem for large firms with advanced ADP departments, but small and medium-sized firms did not possess the capabilities to modify the data to the required format. The recession of the 1990s further decreased the amount of in-house specialized ADP personnel.

Increasingly, firms began to buy standardized software, and thus cooperation with software providers became extremely important. Negotiations with them began in 1991, and by 1993 there were already software packages that were compatible with the ePost Letter system. The cooperation with the providers escalated in 1994 when the PEM unit hired a new employee who had worked in IBM and thus already had the necessary contact network. She describes the beginning of her career at Post as follows:

> I was organizing these training days for software companies in different ePost Letter-center cities and elsewhere, as well. I was on the road then, literally digging up software companies. Digging up applications, what they have got ... [Then] I called and asked if I could come by and tell them [about ePost Letter]. I was literally running from company to company.
>
> (Product manager, Atkos)

Software providers sold the software directly to the customers, which made them an important information channel between the Post and the users of the ePost Letter service. Especially at the beginning, it gave them a competitive edge to be able to provide software that supported the service, and it was only later that it became more of a standard feature. Hence, both the Post and the software providers benefited from the cooperation, although at the same time they were also increasingly competing[14] with each other. The software packages also enabled firms sending less than 5000 letters per batch to use the service.

At first, the customers had used modems to deliver the data to the post, and they seemed to work well with small batches. However, a new data-delivery system was needed in order to cope with bigger volumes. The solution was found in Datanet (data communications network), through which data transfer began in 1992. The first to experiment with big volumes was the PT Finland Group itself. They decided to start sending their pay slips (around 25,000 letters per month) as ePost Letters and delivered the data via Datanet. All this ran smoothly, and Finland Post began to sell the Datanet opportunity to outside customers. Hence, Datanet in particular increased the possibilities of the biggest letter senders to take full advantage of the service.

Thus it was both the software providers and the new data-communication network that enabled the creation of the infrastructure needed to provide the service to a larger number of customers. The next task was to interest the customers in the service.

Increasing interest in ePost Letter services

The Finland Post selling network took care of selling the service, and the PEM unit took responsibility for the form design and provided technical support for the customers. Hence, it was mainly the in-house sales force (around 300 people) who did the selling on a personal level. Major customers were either visited personally or were invited to come and see the ePost Letter processing on site. However, these employees were not used to selling this kind of service, which caused them anxiety:

> They were more or less afraid of the ePost Letter. They said they didn't dare discuss it with clients because "if they ask something technical, I don't know what to tell them". That was a really big issue for our sales people.
>
> (Customer relationship director)

Thus, at first the customer relationship director visited customers with the sales personnel with a view to teaching them to sell the service at the same time. This proved to be too time-consuming, however, and inefficient in terms of reaching large numbers of customers. As a consequence, a team of ten persons was hired to work in sales support. The team began by producing sales-support material (brochures and transparencies, for example), and then they started to train the sales organization, focusing particularly on customer benefits. As the training proceeded, the sales force developed trust in the new system, and their self-esteem grew: they were selling the Post's new innovation, a service based on advanced new technologies, from a position of confidence.

The main selling points were cost-effectiveness and the simplicity and confidentiality of the new system. Concretizing these benefits was considered rather easy: for instance, *cost-effectiveness* was sold on the basis of detailed calculations:

> Then we presented them with facts and calculations, where we'd calculated the value of the job if you did it yourself, as in how much it costs if a secretary first types the letter, then prints it, buys an envelope and a stamp and puts the letter in an envelope. We got a very impressive calculation from that, which showed that the price of an ePost Letter was much less.
>
> (Director, Letter and Direct Marketing Services)

Similarly, it was emphasized that the system was *easy to use*:

> The brochures had a very prominent image of the enter key, as on the keyboard. By pressing enter, you'll send a million letters ... or something like that. In a way it was the ease that came from replacing the manual mailing and the physical work.
>
> (Director, Letter and Direct Marketing Services)

In convincing customers about *confidentiality*, it was emphasized that ePost Letter was subject to both the privacy of correspondence and bank secrecy. As workers in a civil-service department, the employees took an official oath that obliged them to maintain secrecy, and this was also highlighted in the marketing. The interface between the customer and the system was protected by firewalls, user identification and passwords, and when the data was received the system automatically sent a confirmation report back to the sender. Furthermore, every employee handling unenveloped ePost Letters had to sign a concealment agreement, and entry to the ePost Letter–handling premises was extremely restricted. The representatives of the Banking Supervision Office visited the Helsinki ePost Centre and were satisfied with the measures taken to ensure confidentiality. The approval of the Banking Supervision Office was an important selling argument, and as a consequence, various banks and insurance companies started using the service. This further escalated sales.

In general, the reference customers were important, and their names (with their approval) were actively utilized in the marketing and selling. *Image-related selling arguments* were also used, and it was emphasized that, by using the service, firms would give themselves a modern and pioneer image in the eyes of their interest groups.

Personal selling was supported mainly by direct marketing (i.e. targeted marketing letters) and, on a minor scale, trade-fair presentations and advertising. *Customer training* was also considered important, and the manual played the key role in this. Customers were also increasingly calling the PEM unit and ePost centers for advice, and this led to the establishment of an ePost Letter technical-support telephone line in 1992.

Customers gave positive feedback and offered ideas for further development. The sales force conveyed feedback to the PEM unit, which found most of it useful and did its best to carry the ideas through. The worst problems have concerned requirements for colors and the handling of Word documents. Customers want colors in their letters but are not willing to pay the extra costs for it, and special kinds of attachments have been difficult to handle:

> Well, customers have all sorts of demands, we've even had demands that "we want to put a pen in there". [Company X] wants to put a pen inside. We just said that it's just something our enveloping machine can't do. That's where the limit is.
> (Former director of the PEM unit; Former managing director of the IDP)

However, Finland Post tried its best to find solutions even to these problems. For example, in order to help small firms, it strived to achieve compatibility between Word documents and the ePost Letter system by negotiating with Microsoft, but negotiations broke down because of Microsoft's massive royalty requirements and delays in the launch of its new Windows operating systems.

The sales force was also doing its best to meet customer requirements. According to the customer relationship director, this kind of behavior came rather naturally: "There's a lot of creativity in this group. In general, sales people have these kinds of survivor types who always try to find a solution somewhere" (Customer relationship director). Thus, at the launch stage people working in the PEM unit and sales were reacting to the customers' requirements. However, at the same time they were actively influencing the demand and creating awareness among the customers. Thus, their behavior was both *proactive and reactive*.

At this stage, too, the service was marketing itself: customers heard about it through word-of-mouth communication, contacted the Post directly and asked how to get it. The growing demand caused problems on the supply side:

> That was pretty much our problem with marketing in the beginning that we started off with high volumes, really fast. They were in a hurry to get the new service. We couldn't cope with such a small team, we just couldn't.
>
> (Former director of the PEM unit; Former managing director of the IDP)

It was not only in the PEM unit but also in selling and marketing that resources were limited. Funding for marketing and selling was a sensitive issue in Finland Post. In this case, additional funding may have speeded up the launch, and more investment in increasing, and particularly in training, the sales force would have been beneficial.

> The days were really long, we were basically flying – I'm sure it could've been more effective if they had invested at the beginning and hired more people. It was probably a big issue for us as well, that we didn't dare to ask, when we knew times were hard anyway. We preferred to put the money into system development and just tried to deal with it. It was quite a struggle to get even one person.
>
> (Customer relationship director)

Furthermore, there was a need for more funding for marketing activities supporting the selling:

> Maybe we would've made more of the launch. We got sort of cornered, in the right way though, with the right customers, but maybe the growth would've come quicker if we'd shouted about it in the media and sold the idea …. Quite often, when the salespeople went to a company, they had to start from the basics. In a way, selling the idea just took so long that when you got to the implementation, you'd got pretty far.
>
> (Director, Letter and Direct Marketing Services)

Thus, the foothold in the market may have become established more quickly if more financial resources had been invested in the launch. Nevertheless, the ePost Letter service was able to establish a position on the Finnish market in 1995, which was the time when the market segmentation became more systematic and selling efforts became more targeted.

The internationalization of the ePost Letter system

It was mentioned above that Nordic Post became interested in the system at the development stage and in 1991 organized an international conference to promote it. It was after the conference that the Nordic countries started to consider how they should proceed with the development and marketing of the ePost Letter system. Finland Post already had a serviceable system, a ready business concept and a growing customer base. Thus, the Finnish representatives wanted to sell this system to other countries, even though the idea of selling it caused some internal consternation:

> There's a funny story from when we'd almost sealed the deal to sell this system. [N.N.] came to tell me at dinner, "Jarmo, you can't do that. You can't charge other National Posts – they've always worked together. If one of them has something the others want then they'll give it to them. And when the tables are turned, they'll get what they want when the others have got it." I said that we'd got something that others would never have, and I wanted money for it!
>
> (Former director of the PEM unit; Former managing director of the IDP)

Hence, Finland Post offered to sell their ePost Letter system to other national postal operators. Norway and Denmark were willing to buy it, but Sweden was not:

> It was a tough thing for the Swedes that they had to buy something that we'd developed. They had a really tough negotiator, who was of the opinion that we should start with a clean sheet, that everyone would lay down one million marks and then we would hire a consultant who'd start to think about what we were going to do. And I said, we're not going to go with it, we already know what we're going to do and we've already done it – we've got a system ready and you can buy it if you want to.
>
> (Former director of the PEM unit; Former managing director of the IDP)

However, since Norway and Denmark had agreed to buy the Finnish system, Sweden had no option but to comply. Finland Post sold 25 percent of its ownership rights to each of the three Nordic organizations, following which, in January 1992, all four decided to find a private firm, NDP, and sell their ownership rights to it. The objective was to generate profits for the Nordic postal operators by selling the ePost Letter system license to other national bodies. NDP took care of the marketing, and of developing the system further, and provided technical assistance for its international customers. It was headquartered in Copenhagen, and Koivunen became the managing director. Cooperation with the Nordic countries seems to have been fruitful, and the division of tasks between them apparently worked rather well:

> You've probably heard this joke, what are the differences between the Scandinavian nationalities? Well, the differences are that the Finns invent it, the Danes manufacture it, the Swedes market it and the Norwegians buy it. And this concept worked well there.
>
> (Former director of the PEM unit; Former managing director of the IDP)

Encouraged by the system's potential, postal operators in France, Germany and Australia came on board, and NDP became IDP within its first year of existence. Shortly thereafter, postal operators in Italy and the United States joined the Board. Users of the ePost system included the Board members and various licensees including postal operators in Singapore, Iceland and Switzerland. Finland Post played an important consultant role in advising international operators about selling the service to their customers:

> It was a pretty good role we got to play. I can remember being very proud of getting to act as an expert for the Scandinavian postal companies first. We had all these meetings here, they came here, I went through all our support material and then we supplied copies of all of it to them. Basically, they then got the benefit of our work and we didn't realize we should have charged them for it. We charged them for the production system, but this we didn't think of in all the euphoria and enthusiasm – all of us, and I personally, thought it was great to see ePost Letter spread beyond Finland, that it was a Finnish innovation and an entire concept developed by Finns.
>
> (Customer relationship director)

Hence, internationally, the main financial benefit for Finland Post was in selling the ownership rights of the system. Even though the free distribution of the sales-promotion material did not bring in revenue, it may still have been beneficial in generating interest in adopting the system:

> On the other hand, maybe it was good that everyone got it. When they bought the production system, they all received this support material and that made it a lot quicker, because it took us nearly a year and a half to build it from scratch.
>
> (Customer relationship director)

The international operators seem to have been as enthusiastic about selling the service as the Finns had been, and when they started they had similar experiences (e.g. involving the sales force). What they learned from the Finnish experiences somewhat alleviated their problems, however.

Consequently, even though Finland Post sold the ownership of the ePost Letter system at the beginning of the launch stage, it seems to have *influenced* the international customers. The scope of this influence was *polycentric*, since the system was sold to countries with the most advanced postal systems. A polycentric approach was also required because the customers were other national operators with their own traditional ways of serving their own customers.

IDP was very successful at first, and the ePost Letter system was licensed to various countries. However, problems started to emerge in the second half of the 1990s, and international expansion ceased. The problems were multifaceted, but they can be classified as development-related, decision-making-related and market-related.

Further *development* of the system was getting more and more difficult, since those involved no longer had direct contact with the customers.

> We're reliant on second- or third-level information. We're still in a situation where we only look at the market and the customer demand in Finland. And the assumption that the Nordic countries are similar is wrong. So, in a way, we have nothing.
>
> (Development manager)

Thus, developers were getting different technical demands from various national operators without knowing the reasons behind them. It was also becoming very difficult to take into account the various requirements of particular national operators and at the same time to develop one system that would be compatible with them all. Consequently, the development proceeded slowly.

Furthermore, following the establishment of IDP, cooperation with IBM became more difficult since the other shareholders thought that the system should not be limited to IBM operating systems. The idea of having an open system was good in practice, since it corresponded to customer requirements. IDP thus made an agreement with an Italian firm, Elsaq-Bailey, who promised to develop a Unix-based system. The development took three or four years and resulted in substantial additional costs. Elsaq-Bailey had no previous knowledge about this kind of system, and IDP could not specify what they really wanted. The development took so long that the system was already technologically old when it was launched, and various operators, including Finland Post, started working on it independently.

The *decision-making* in IDP was getting more complicated along with the expansion of the Board as the new shareholders joined it. It became extremely difficult to achieve consensus, and the situation was made more difficult because the shareholders were also the customers. This put pressure on the pricing policy, for instance.

Market-related difficulties arose from the restricted number of potential customers. IDP had agreed that old shareholders and licensees were entitled to get the further developed systems partly for free, but due to the limited number of advanced national postal systems, suddenly there were no new licensees, and development funding began to dry up. Furthermore, since the license fee was tied to the GNP of the country, and since the new licensees increasingly came from countries with smaller GNPs, they brought in less and less revenue.

The organization and philosophy of IDP was re-thought several times in the second half of the 1990s. It was acquired by the Norwegian Ergo Group, owned by the national postal operator, in November 2001, and its name was changed to ErgoIDP. Currently, the firm is marketing its hybrid mail solutions to postal administrators, and increasingly to companies.

In hindsight, it was acknowledged that perhaps instead of striving to internationalize the system together with other operators, it may have been more beneficial for Finland Post to have tried to go international alone.

We should've probably had more self-confidence at that time, and we should've dared to invest in it. Hive it off to a subsidiary and develop it a step further. [...] That we engaged in discussion with the postal operators of these countries ... We constantly had to make compromises and it didn't really work out.

(Director, Letter and Direct Marketing Services)

On the other hand, going it alone would have been extremely difficult, since Finland would perhaps have been too small a reference point. The attitudes of the Paris Group toward the idea in the 1980s further support this kind of reasoning.

The demand for the ePost Letter service has been growing steadily. Since the launch proved to be successful, people were encouraged to think about the next step. Koivunen returned to Finland in 1995 and formed a small team to consider future communications solutions.

Jarmo was really excited about this. We were there at the head office, we had a few rooms and we were all in the same close quarters. There we were then, drawing and brainstorming these electronic postal solutions ... We had drawn some pretty impressive doodles that this is what we're going to make ... And then what happened was that the management at the Post at the time questioned our business plans and time schedules.

(Customer relationship director)

Thus, the ideation of the new systems was pegged, and the senior management urged closer cooperation with the telecommunications unit, or even total absorption. This decision caused frustration among the development team, especially when the telecommunications unit was privatized at the beginning of 1997 and, due to the former decision, gained developmental advantage.

The future of the ePost Letter service

Organizational changes influencing the development of ePost Letter followed one after another. The two biggest players in the market, Finland Post and Tieto-Enator, founded a joint venture, Nordic Printmail Ltd, in 1998, and the new firm became a clear market leader in printing services. Since then Finland Post has gradually increased its data and material-flow management services, and it has become a major provider of intelligent logistics services in the Baltic Sea region. This growth has taken place both organically and through acquisitions.

In assessing the impact of ePost Letter, it is worth emphasizing that it also brought along new service ideas such as eCard (a card sent the same way as an ePost Letter), an electronic archiving service and Multiletter (the company delivers the electronic material to the Post, which then delivers the letters to consumers in the format of their choice – either on paper or electronically). Nowadays the ePost Letter service plays an important role in the *revenue generation* of Finland Post. In the future, the development of electronic communications is likely to reduce the demand for the service, although the change will probably be slow because some of the receiving customers will not be ready to switch to electronic reception: "At least we believe strongly that paper will never completely disappear. Paper is such a good interface that there will always be a demand, almost endlessly .." (Director, Letter and Direct Marketing Services). The ePost Letter service could be also seen as a *facilitator*, with an important role in involving customers in the new electronic mail solution offered by Finland Post:

> To put it nicely, it's more like ePost Letter is a bridge to the solutions of the electronic world of the future. It's really the message we give our business customers ... that when you're with us, and the technology evolves, you'll get the best solutions there are on the market, and that we're actively working to achieve this.
>
> (Customer relationship director)

However, it is acknowledged that competition on the electronic-mail service market is very tough, and it is not likely that customers of the ePost Letter service will automatically choose the Post as their provider. Although Finland Post has at least one advantage over its competitors in that it has very long experience and thorough knowledge of communication behavior and communication flows in Finland, whether or not it is able to benefit from this knowledge remains to be seen.

5.3.7 Customer-related proactiveness during the ePost Letter development process

In sum, it took 13 years for ePost Letter to change from an idea to an established innovation. The different stages of the development process are illustrated in Figure 5.13.

Customer-related proactiveness seems to have differed at different stages of the development process (see Figure 5.14). *Since the ePost Letter has been sold*

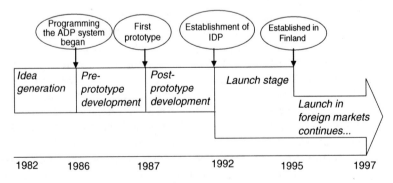

Figure 5.13 The innovation development stages of ePost Letter.

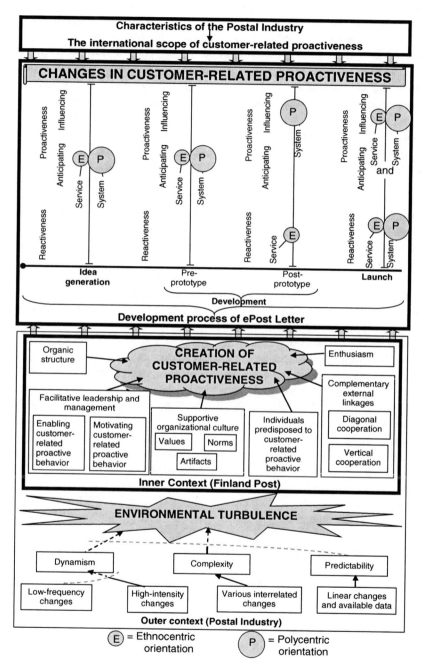

Figure 5.14 The change in customer-related proactiveness during the development of ePost Letter.

both as a service and as a system, both are illustrated in the figure. The inherent characteristics of the postal-service industry have influenced the international scope of customer-related proactiveness in both the service and the system development. The scope of the former has been ethnocentric and restricted to the Finnish market, whereas the system development has been polycentric, i.e. oriented toward certain foreign markets with developed postal systems. A geocentric approach was not feasible due to the limited number of such countries. A regional approach was not possible either, since the long postal traditions in individual countries and the fact that each country had one customer, the national post operator, favored treating each country individually.

At the idea-generation stage, the market opportunity was mainly *anticipated.* Finland Post knew its biggest customers rather well and assumed that there was a latent need for an ePost Letter service. Furthermore, close contacts with foreign postal operators strengthened the presentiment that there would be an international market for the system.

Before the prototype was developed, Finland Post was still relying on its good knowledge and *anticipated* the needs of its customers, and it was only afterwards that it became possible to *react* to them. At the same time, it started actively to influence prospective international customers. Since most foreign national operators were not very interested in hybrid services, it was necessary to actively promote awareness of the system.

Behavior at the launch stage seems to have been both *proactive and reactive.* Finland Post was actively creating awareness among its Finnish customers and was also reacting to the further requirements of its service customers. The international launch of the system took place through NDP, which also took care of developing it further. It tried to react to the various requirements of different national operators and, at the same time, to create demand for the service by influencing them.

When the ePost Letter was developed the turbulence in the environment of the postal-service industry was rather low, and whatever significant and interrelated changes were taking place they were slow and linear. The fact that there were data available on the coming changes also contributed to the predictability that decreased the environmental turbulence and increased Finland Post's preparedness.

Although the organization of Finland Post was rather mechanistic in general, there was room for another kind of behavior. The team that developed the ePost Letter service was able to establish a sub-organization that seems to have been very organic and furthermore had an organizational culture that supported proactive behavior toward the customers. Leadership in this sub-organization was both enabling and motivating as far as this kind of behavior was concerned. However, against the theoretical framework, the building up of customer-related proactive capabilities was not particularly emphasized in this case.

Individuals working in the organization seem to have been predisposed to customer-related proactive behavior. Complementary external linkages apparently also played a role in the development and launch of ePost Letter. The cooperation with IBM was important at the development stage, although the aim was to achieve technical solutions and not to foster customer-related proactiveness. At the launch stage, however, proactiveness toward customers was further enhanced through cooperation with software houses in Finland and with national postal operators abroad. Hence, in Finland the cooperation was diagonal, but abroad it was vertical, since the foreign operators were, in practice, regarded as customers of the ePost Letter system.

Furthermore, even though the interviewees said that more finance at the launch stage would probably have speeded things up and have enabled larger numbers of customers to be dealt with, it still may not have increased proactive behavior toward them. There does seem to have been enough finance at the development stage, however, thus the role of liquid financial assets was not particularly accentuated in increasing proactiveness in this case.

In addition to the issues presented in the preliminary framework, the *enthusiasm* of the workers was prevalent during the development of ePost Letter, and it even seems to have been conveyed from the developers via the sales force to the customers. It apparently increased the propensity of the workers to influence the customers and is therefore added to the framework.

5.4 Hi-Fog

5.4.1 Hi-Fog as a radical innovation

Hi-Fog is a fire-extinguishing system. Fire-extinguishing sprinkler systems existed in the nineteenth century, when water was the natural extinguishing agent. Carbon dioxide (CO_2) systems emerged in the 1940s and were used mainly in the extinction of flammable liquid and gas fires. At the same time, steam-flooding systems were used for specific applications, such as in lumber-drying kilns, before they were replaced by halon in the 1960s. Thus, the idea of using water mist to extinguish fires is old. Marioff Corporation was the first company to succeed in turning this idea into a commercially successful system that complies with the fire-protection level of today. The Hi-Fog system (see Figure 5.15), which uses water mist, was developed in 1991.

Hi-Fog is a *commercial success*. Its extinction capacity has improved over the years, and new applications have been developed. Because of the small drop size and minimal water usage, Hi-Fog systems are being used in a very wide range of applications. Hi-Fog was originally developed as an alternative to conventional sprinkler systems on passenger ships. Marine applications currently account for almost half of the company's turnover. The key focal areas in the marine business are passenger-vessel retrofit projects, cargo vessels and naval fleets, and Hi-Fog currently has a world market share of over 90 percent of water-fog installations in cruise and passenger vessels. However, there is a growing demand for onshore applications in areas in which minimal water damage and easy installation produce major benefits to customers, such as underground trains and stations, industrial buildings, computer and telecom

The Hi-Fog Sprinkler System consists of a number of Hi-Fog sprinklers connected by fresh-water-filled, stainless-steel small-bore tubing to a pump unit via section valves.

Figure 5.15 The Hi-Fog system.

rooms, gas-turbine installations, hotels and museums. The volumes for onshore systems are substantially higher than those for marine systems, and Hi-Fog's market share in the onshore business is growing steadily: it accounted for over 50 percent of the turnover in 2006.

The *technological novelty* in Hi-Fog originates from the innovative combination of hydraulics know-how and extinction techniques: it combines the right drop size, drop-size distribution and high velocity. Small drops effectively absorb the fire energy, cool the surrounding hot air and gases, and prevent oxygen from mixing freely in the combustion area. The mist (average drop size of 50–120 μm) is delivered through patent-protected[15] Hi-Fog sprinklers or spray heads, and the water-mist system incorporates considerably greater pressure difference than conventional water-sprinkler systems. Pressure of below 140 bars is achieved by using high-pressure pumps or charged gas cylinders, and the high pressure allows the mist to penetrate the hot flue gases and reach the source of the combustion.

Compared with previous extinction systems, Hi-Fog provided certain *substantial benefits* for its adopters. These benefits are basically related to its fire-suppression capability, the small water amount, ease of installation and friendliness to the environment. Fire-suppression capability refers to the speed of extinction and the minimization of smoke damage: quick extinction is due to the three-dimensional effect of water mist (conventional sprinkler water only takes effect along two dimensions), and the smoke-absorption capability is higher than in former lower-pressure systems. Furthermore, the water mist cleans the combustion gases. These are important benefits, since the problems caused by smoke and gases are often more dangerous and noxious than the flames themselves. Furthermore, water mist is rather effective in absorbing radiant heat, thus keeping heat levels below what is attained with a conventional sprinkler extinguisher.

In extinguishing a fire, the Hi-fog system utilizes only 10 percent of the water amount needed by a conventional sprinkler system. Consequently, water damage is much reduced and clean-up costs are lower. Moreover, large amounts of water may affect the stability of a vessel: a water-mist extinguisher system weighs almost 90 percent less than a conventional sprinkler system. This is particularly important for marine installations. Furthermore, the system is easier and faster to install because it only needs a small pipeline circuit. Thus, it can cut down installation time by a factor of between four and five to one, which is of significance in retrofit projects when time is at a premium. Water mist is also environmentally friendly since it contains no chemicals, only fresh water. This is clearly a substantial benefit, since many former systems used environment-damaging halon or CO_2. Water-mist extinction can be used safely even when people are exposed to it, unlike CO_2.

Hi-Fog has also *changed the behavior* of its adopters. Above all, it has enabled the installation of extinction systems in many vessels that would not have tolerated the heavy weight of conventional systems. It also requires minimal downtime: for instance, retrofits for vessels have been installed by riding crews while the vessel was in service. The emergence of Hi-Fog has also influenced the creation of standards and regulations for fire-extinguishing systems, and it seems to have revolutionized the thinking in the fire-protection field, in the sense that old conventional systems are more and more challenged to prove their fire-extinguishing capabilities. "Lately, there's been a trend for updating everywhere, a general process of putting systems on the same line. [...] That's why they [conventional systems] are under a microscope and people realize that they're not as good as they'd imagined" (Manager, Research and Development). Besides its innovativeness, there are also certain factors inherent in the system itself that seem to have influenced its development and marketing. First, the systems are *custom-made*, and the installation may be spread out over two or more years. Usually they are first sold and then designed and installed to fit the customer's premises. Second, the systems are *expensive*. On average, one delivered system is worth 1.7 million euros. Third, although fire-extinguishing systems play *a pivotal role in saving lives and property*, their functioning capability is evident only in real accidents. This means that buying decisions are largely based on trust and the seller's ability to convince the prospective buyer that the system will function properly in the case of fire. Fourth, the fire-protection business is tightly *regulated* and thus extinction systems need to comply with certain standards.

Client characteristics have also influenced the way in which the systems are sold. Those in the shipping industry comprise approximately 50 big shipowners and docks, and consequently, the business relationships are long term and based on trust: "This is very much based on knowing the people, and that they either trust you or don't, right? You just take the order and you don't really need that much paper" (CEO). On the other hand, the clients in the onshore business are a very large and heterogeneous group (e.g. hotels, museums, supermarkets, telecommunications companies, power stations and traffic enterprises). Selling to them is usually transaction-based, which makes it very different from selling to the shipping industry.

5.4.2 Marioff Corporation – the developer of Hi-Fog

Marioff was founded in 1985 in Vantaa. The firm began by providing specialized hydraulics services and products, mainly to marine and offshore oil-and-gas markets. The combination of water-mist technology and hydraulics forms the base of its current business.

Marioff's founder, CEO and Chairman of the Board, Göran Sundholm, was the key person in the invention, development and launch of Hi-Fog. He invented the system and participated actively in its development and marketing. He has been described as "the father of Marioff"[16] (Puustinen 2001)[17] and as "the heart and soul of Hi-Fog" (Manager, Research and Development). According to the R&D Manager, the main reasons for Hi-Fog's success were its superior product quality and the CEO's persistence. The CEO calls himself both market- and R&D-oriented and says that he enjoys both inventing and selling. "If you haven't sold it, there's nothing to develop. First you have to sell, and then develop. Isn't that how it usually goes?" (CEO). He seems to be an innovative entrepreneur, who filed his first patent application at 17 years of age, founded his first company at 26 and has over a 1000 international patents and patent applications. He has a technical education and has acquired his know-how mainly through practice. Finding solutions to clients' problems or technical complications is so rewarding that it keeps him working long days, even at weekends. He has been rewarded for his innovativeness. In 2002, he received the Finnish Engineering Award for his pioneering work as a developer of novel protection technology, and later in the same year, he received the Finnish National Board of Patents and Registration Award for his work as an inventor: he had more patents to his credit than any other person in Finland. Most of these patents are related to Hi-Fog technology. He won the national Ernst & Young Entrepreneur of the Year® award in 2005. The CEO thus clearly seems to have *a proactive, customer-oriented personality*. Since his influence in Marioff is strong, it can be assumed that he and his example have contributed to the proactiveness of the firm as a whole.

In 1991 when Hi-Fog was created, Marioff employed 14 people. Although there was no explicit strategy, the firm was aiming at annual growth. According to the CEO, times were so hectic that there was no time to think about strategies. The organization was very informal: the R&D Manager stated that "there was no organization" at that time.

The organizational structure was rather *organic*, inter-firm communication was open and emphasis was clearly on "getting things done" rather than on "formally laid down procedures." However, the decision-making was *autocratic* and firmly in the hands of the CEO, as he described it: "It was quite easy, I decided everything." The CEO further stressed the fact that, generally, democratic decision-making tends to be destructive for an enterprise.

Furthermore, the organizational culture in Marioff emphasized *fast decision-making*. Since the firm was very dynamic and decisions were made quickly, the employees had to be flexible and to tolerate stress. Those who managed to fulfill these requirements still work for the company, those who did not soon left. An example of the style of decision-making and pace of working is the installation of an extinguishing system in an Asian cruise vessel. The tender was invited one Thursday and accepted on Friday of the same week. On Saturday, the vessel left Turku with the equipment and assemblers on board. The whole payment (millions of euros) was received on the following Monday, and the system was installed during the voyage to the Far East. Furthermore, all these transactions took place without a written order.

Humor seems to have softened the pace and pressure that was prevalent in the firm. On various occasions the interviewees emphasized that working has to be fun: "I've always said that if business gets too serious, then you should finish it. You've got to have a bit of fun in it, as with working in general, it shouldn't be all that serious" (CEO). In selecting employees, the emphasis was on flexibility and the toleration of stress. Proactive capabilities were not focused on the selection or on the training, and the *building of proactive capabilities* was thus not emphasized in the firm. Training, in general, has been rather limited. At first it took place through practice, and new employees learned along with the system development and the fire tests. However, training has become more formal and systematic in recent years as the firm has grown.

Enabling customer-related proactive behavior was not very active either. Knowledge was not openly shared between departments, and over the years, the constant research work together with the increasing pressures of secrecy have impeded cooperation between marketing and R&D. The system development was so dynamic in the 1990s that there were practically no brochures, and the training of the sales personnel was rather exiguous. Consequently, this strained the R&D department, since the sellers and installers in the field were constantly calling them for advice. However, since 2000 the marketing and sales personnel have been better informed about the products. According to the Manager (R&D), this change took place due to the emergence of various standardized products and systems. Nowadays the sales personnel is regularly trained and supplied with updated brochures.

The CEO of Marioff seems to have emphasized *motivating customer-related proactive behavior*. Energizing, envisioning and inspiring seem to describe him, and he has demonstrated his own enthusiasm throughout the Hi-Fog development process. He has constantly aimed to seek out new opportunities and has tolerated failure without ignoring it. Nevertheless, the extent of customer-related proactive behavior could have been even bigger if successful work had been particularly rewarded, which was not usually the case. The interviewees described the management's reactions to employee success as follows: "Then you didn't get a scolding" (CEO). "Yes, that's really true!" (Manager, Research and Development). Another feature in the firm has been the strong emphasis on continuous *research and development*: product development, full-scale fire tests and patent protection have been prioritized in the money allocation. According to the CEO, Marioff has probably spent more money on research and fire testing in the last decade than any other company in the fire-protection community. In the early 1990s, over 10 percent of the turnover was invested back in R&D, and a substantial part of that was spent on thorough patent protection. Consequently, the firm has been the recipient of several major national and international awards, including the Safety at Sea Award (1992), the Innosuomi Award (1995) and the Finnish Technology Export Award (1998). In order to protect the value of their R&D investment, Marioff has striven to keep design and manufacturing in-house.

The pioneer role of Marioff as a developer of Hi-Fog and its active influence on standards have been strongly promoted in the firm's communications and were also emphasized by the interviewees. It thus seems that proactive behavior is valued rather widely throughout the organization, indicating *the intensity and crystallization* of the proactive culture in general. The firm's small size at the beginning and the CEO's proactive personality have probably contributed to this cultural norm.

Fast *growth* has been characteristic of the company since the launch of Hi-Fog. When it was developed in 1991, the turnover was around FIM 4 million, and the firm employed 14 people: in 2006, the turnover was over 93 million euros and the number of employees was 347 (see Figure 5.16). The company changed from a limited partnership to a limited company in 1994, and the name was changed to Marioff Corporation in 1999. According to the R&D Manager, the growth has changed the organization: "Well, I'd say that it was around the year 2000 when things started to get more ordinary. It started to become an ordinary, boring company. It seems that with growth comes bureaucracy" (Manager, Research and Development).

The big change that seems to have accelerated this development toward being "a normal firm" took place in 2001, when a private equity firm Nordic Capital acquired 50 percent of the company from the CEO. According to him, the motive for selling was to enable continuing growth by acquiring additional capital and an experienced business partner. The acquisition brought changes in the management: a new president was appointed, the CEO became chairman of the board, and two businessmen from outside the firm were put on the Board. In May 2007, Marioff's shareholders accepted UTC Fire & Security's offer for the company. It is expected that the new American owner will provide a stronger platform to ensure Marioff's continued global growth.

Although the growth seems to be making the firm more mechanistic in its organizational structure, it could still be characterized as more organic than mechanistic. In other words, the emphasis is still more on getting things done, control is informal and on-job behavior is flexible. Organizational factors are

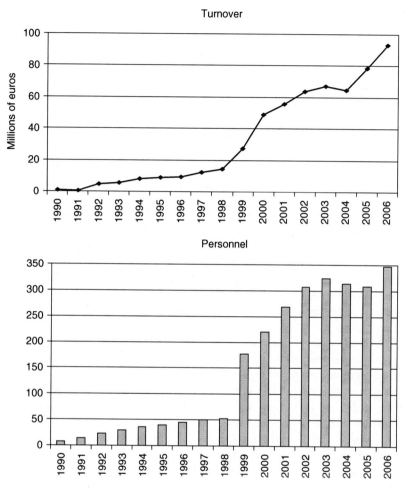

Figure 5.16 Turnover and personnel of Marioff 1990–2006 (Sorsa, e-mails April 4, 2003 and June 14, 2007).

discussed further in the following sections, alongside the description of the development of Hi-Fog.

5.4.3 Environmental turbulence

The fire-protection industry consists of organizations specializing in the design, supply, installation and maintenance of a broad spectrum of technologies and products ranging from fire detection and warning systems to extinguishing systems. It seems to have been rather conservative and prejudiced against new technologies and procedures. At the end of the 1980s, the industry was rather *stable*: old systems such as conventional water sprinklers and halon systems were prevalent, and traditional players shared the market.

The existing supplier–buyer relationships generally make it difficult for new suppliers to enter the field. However, it was particularly *difficult for any new player to enter* the fire-protection market with a new system, because the industry was and remains highly controlled. Any new commercial products or systems are required to go through extensive and expensive series of fire tests, and *regulations* play a major role in the extinction business. Since extinction systems are considerable investments, customers are not usually willing to acquire them voluntarily.

Although the shipping industry is regulated through national legislation, in practice this is adapted to international treaties pertaining to shipping and navigation. Internationally, the key organization influencing regulations is the International Maritime Organization[18] (IMO). Because the shipping industry is highly global, the IMO promotes the adoption of international conventions and universal standards. In practice, governmental institutions, such as the Finnish Maritime Administration, ensure that the shipping industry complies with the standards and legislation. Firms can influence the regulations, but the process is slow and usually takes between five and ten years.

Whereas regulations and standards concerning the shipping industry are global and uniform, onshore regulations are highly heterogeneous and country-based. National laws regulate onshore business, but in practice various insurance companies further restrict it. They require that the systems installed in their insurants' premises comply with certain standards.

Although, as mentioned above, the fire-protection industry was stable, there were major regulatory changes coming up that influenced all firms providing fire extinguishers. The first one was related to the growing emphasis on *environmental protection*. Halon chemicals that were largely used to extinguish fires were proved to deplete the ozone layer relatively quickly. Thus, in 1987, most countries adopted the United Nations Montreal Protocol, according to which the industrialized countries agreed to phase out halon in 1994, while the deadline for developing nations was set at 2002. This put pressure on companies to develop alternative extinction systems.

The second change was related to the *safety regulations*. In April 1990, a fire on board the *Scandinavian Star* caused the deaths of 158 people. This highlighted the need for the IMO to take urgent action to bring into force new international rules for the shipping industry. Thus, it was already clear to the shipping companies in 1990 that sprinklers would soon be compulsory in passenger vessels. In 1992, the IMO decreed that all new and existing passenger vessels must be fitted with automatic sprinkler systems by 2005. However, traditional systems were not suitable for many of them because the considerable weight would have compromised the vessel's stability.

In sum, in the late 1980s and early 1990s, the regulatory environment of the fire-protection industry underwent major changes that motivated firms to develop completely different extinguishing systems. Accordingly, *the dynamism*

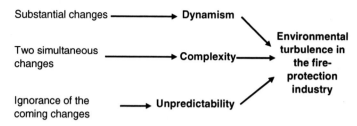

Figure 5.17 Factors contributing to the environmental turbulence in the fire-protection
industry.

in the environment was high in terms of the intensity of the changes (see Figure
5.17).

Furthermore, there were two big changes in the environment that were related
to each other in the sense that they both required new extinction systems to be
developed. Consequently, the *complexity* in the environment also increased.

These changes could probably have been predicted beforehand, since there
was a trend toward tightening safety regulations in vessels in the early 1970s,
and environmental concerns were growing in the early 1980s. However, it seems
that neither the shipping companies nor the fire-fighting industry anticipated the
coming changes, since they took no measures to develop new extinction
systems. They were thus ignored, which, according to the theory, creates *unpredictability*. Dynamism, complexity and unpredictability together constitute
environmental turbulence.

5.4.4 Idea generation

The development of Hi-Fog started in January 1991, when the manager of the
Swedish shipping company Euroway noticed that the conventional extinction
systems were too heavy for their two vessels[19] that were under construction.
Since he was also aware of the coming IMO regulations requiring sprinklers
onboard, he was clearly in trouble. He asked the Marioff CEO to "Do something" and gave him 40 percent of the final price as a down payment for an
equivalent but lighter system, setting a time limit of half a year. Thus, the situation for Marioff was advantageous in the sense that it had received a significant
order for the system and some finance for its development.

Money was certainly needed, since the fire tests were very expensive. External finance was not sought, since it would have been difficult if not impossible
to get at that stage because there was not even an invention, just an undefined
order from a client. However, in terms of liquid financial assets, the situation
was favorable. Before founding Marioff, the CEO had made substantial sums by
selling his previous firm, and he was able to invest almost seven million euros in
the development of Hi-Fog.

Thus, he had to take a considerable personal financial risk in order to develop something new, but he was also confident of success. According to the R&D Manager, he simply does not understand failure and always manages to work things out. As the CEO said, they had to succeed and this imperative was the best motivation. He admitted that if they had not had an order, they would probably have given up.

> We didn't have a clue what we were promising, and luckily so. I don't think we would've promised anything otherwise. Because, the way it usually goes is that you go and sell something and then come home thinking what on earth did we sell them? And then you're in a situation where you have to invent something to solve the problem.
>
> (CEO)

Even though Marioff employees had strong hydraulics and piping expertise, they knew very little about fire control, as the following statement by the CEO illustrates: "I once witnessed a forest fire when I was about ten years old. That was my experience" (CEO). Later, they admitted that this was clearly an advantage, since they approached fire extinction from an unconventional perspective, and used high-pressure hydraulics technology that was strange to the fire-protection industry, although very familiar to Marioff. Thus, experience would have been a hindrance in terms of finding a novel solution to an old problem. According to the CEO, the financial assets and the personnel resources were sufficient at this stage. If they had had more, it may have inhibited their novel approach to fire extinguishing: "If I'd had more money, then I might have gotten some crazy idea to hire some expert and probably it would have failed" (CEO). It was clear that the customer demand for a lighter water-based extinction system was realizable only by diminishing the amount of water, and the search for a suitable system was very intensive. According to the Manager (R&D), the role of serendipity was not very significant in this, since the number of tests was considerable.

Tests were conducted in January at SP (the Swedish National Testing and Research Institute) in Borås. In the first series, 300 mattresses were burnt, with poor results. However, the CEO considers these tests successful since they gave an insight into what not to do. Furthermore, they all learned how fire actually works. The CEO returned to Finland to improve on the idea and established that they had to reduce the drop size and increase its velocity. New tests were conducted in Borås, and this time the system worked. Thus, *the idea of Hi-Fog was born*.

In light of the theoretical background, the market opportunity was *anticipated*. There was an undefined request, i.e. a problem statement, from the client. Marioff used applied research in order to find an answer to the client's problem. The search was systematic, and as a result, an idea was generated inside the firm.

At the time Hi-Fog was created, people at Marioff were rather *enthusiastic* about the new development work. Their enthusiasm stemmed from the small size of the firm and its clear objective to develop a new system that would fulfill the client's needs. Furthermore, since advance payment had been received from the client and most of the firm's (more specifically the CEO's) money had been spent on the trials, success was vital. The enthusiasm was manifested in the trials that were being conducted day and night.

> In SP [the Swedish National Testing and Research Institute, where the Hi-Fog trials were conducted] they said: 'There are three kinds of working hours: regular hours, overtime hours and then … Marioff's hours.' We worked day and night there, too.

> (CEO)

At this stage, the emphasis was on fulfilling the promised order, and there was no indication of the coming big success of the system. However, because of the international nature of the shipping industry, it was assumed that the potential market for water-mist extinction systems would be global. According to the CEO, the orientation always has to be global when clients come from the shipping industry. Thus, the orientation of the firm was *geocentric* at all the stages, from idea generation to the launch and beyond.

5.4.5 Development

The prototype of the water-mist sprinkler system was already available in January 1991, when a series of experimental fire tests confirmed that such a sprinkler, using considerably less water than traditional sprinklers, suppressed and controlled simulated cabin fires. The fast emergence of the prototype supports the findings of Veryzer (1998a), suggesting that radical innovations require rapid development. In the case of Hi-Fog, the prototype was developed in parallel with the emergence of an idea concerning a new system. Hence, *the ideation and the pre-prototype development phases could be considered one phase.* This convergence may have resulted from the fact that the whole innovation emerged in the process of fulfilling an existing, although undefined, order.

The post-prototype development stage ranged from late January to April 1991. Thus, although the development has been a continuous process, for the purposes of this study, it is limited to the time between the idea and the launch of Hi-Fog. It was a very brief but intensive stage and involved constant overtime working.

Hi-Fog was developed in cooperation with The Swedish National Testing and Research Institute (SP) and The Technical Research Center of Finland (VTT). Their role was to provide facilities in which fire tests could take place. According to the R&D Manager, who was then working for VTT, Marioff was a rather unusual client. It was a small, unknown firm with particularly strong self-assurance, carrying out extensive tests. The intensive work pace also caused some resentment among the VTT personnel, who were not fond of working at weekends and late at night. On the other hand, having Marioff's people around was also considered positive and fun. The researchers at VTT in particular were enthusiastic about the innovative and pioneer spirit that was apparent.

Product development was very *open* from the very beginning, which illustrates how enthusiastic the developers were. "Then it felt like every passer-by was invited to see the tests" (Manager, Research and Development). "Yes: here's the fire. Let's see how the system functions!" (CEO). According to the R&D Manager, this was perhaps one of the biggest mistakes made during the development phase in that it also revealed to outsiders the unsuccessful experiments along the way, and it later proved to be very difficult to change their impressions about the functionality of the system. The openness also indicates that, at that time, there was no indication of the coming success of the system – indeed the CEO acknowledges that they did not pay much attention to the competition. According to the theory, secrecy concerns are normally prevalent in the development of radical innovations. However, this case illustrates that openness may prevail since at this point neither the firm itself nor the competitors understood the breakthrough potential of the new product/system.

No market research was conducted before the launch, as that was not the custom in the firm. "The only thing we ask is: 'Where is the order?' We don't waste time on market research. We have always locked into an order right off" (CEO). The *lack of market research* may partly stem from the characteristics of system selling, in which complex combinations of products and services are provided to the customer as an answer to a specific problem (Dunn and Thomas 1986). However, the literature has emphasized that, even in this situation, system-selling opportunities can be scanned and thus anticipated (cf. Cova *et al.* 1994).

Since the client had solicited an order, the development work was focused on fulfilling it. Although the client did not directly participate, its desires influenced the development of an even lighter system. Thus, it could be concluded that, at this stage, the firm *reacted to* the needs of the customer. Although there was only one customer at that time, its needs stemmed from global changes in fire-protection requirements in the shipping industry. Thus, the reaction had an implicit *geocentric* scope.

5.4.6 Creating the market

It was extremely important for the firm to get the system onto the market quickly and thus to attract more orders. Hi-Fog was introduced in April 1991 at the Cruise and Ferry exhibition in London, and by 1995 it had achieved an established foothold in the marine business. The launch is still under way in the onshore business.

Assuring market acceptance

The Cruise and Ferry exhibition seemed the right venue for the launch, since it was necessary to explain the benefits of Hi-Fog to the key decision-makers in the passenger-ship business face to face. In order to highlight the benefits of the system, a *demonstration* involving the extinguishing of real fires was organized every hour in the car park of the exhibition area. Although the exhibition officials were not very happy to have the fire-test container in their parking lot, the ship owners appreciated seeing the extinction of real fires. It is worth noting that, since then, real fire demonstrations have played an important role in the marketing of Hi-Fog. The fires are big so as to make the demonstration as colorful and as impressive as possible. Nowadays, customers and authorities are often invited to Marioff's own premises to see the demonstrations, which have also been effective in accentuating the differences between Hi-Fog and conventional sprinkler systems.

> What's made this journey easier is that these fire tests have also been conducted with existing and approved systems and they don't work the way people think they do. They're not as good as everyone assumes they are, because they've never seen them. It gives a frame of reference to compare different systems in.
>
> (Manager, Research and Development)

The comparisons between different systems have also been emphasized generally in communications to customers, and particular stress has been put on fire-protection capability and safety during accidental discharge. Throughout the launch, international *exhibitions* were a fertile ground for making new contacts, showing off the latest Hi-Fog applications and strengthening existing business relationships. The firm has participated actively in both fire and safety and marine exhibitions.

The best way of raising awareness about the new system among clients in the marine business has been by *visiting* them. After all, it comprises only around 50 shipping companies, and their people already knew the CEO of Marioff as a provider of hydraulics services and products. Consequently, they also seemed to trust the company and its products.

Immediately after the launch, the main emphasis in the sales pitch was on the lightness of the system. The small amount of water and the smaller risk of water damage were also accentuated. Once awareness had been raised in the shipping industry, the emphasis shifted to maintaining business relations.

References are highly valued in the marine business, since investments are considerable. No shipping company wants to be the first, but in the case of Hi-Fog, the first client with its order for two systems was already there. This existing order was strongly emphasized in the marketing, and after the Cruise and Ferry exhibition, it was followed by an order for a system for a new ferry and for the first retrofit systems on two other ferries. Retrofit installations emerged as a consequence of the anticipated tightening of IMO regulations, and they proved to be rather lucrative, since they could be sold and installed without the years of lead-time that usually applied to new vessels.[20] This was especially important for a small firm like Marioff. It was also advantageous to gain a foothold in the retrofit market, since only a very limited number of new vessels were built annually. All in all, references have been used extensively in the publicity and include extensive lists of previous Hi-Fog clients and of fires that have been extinguished by the Hi-Fog system. Comments from clients about their experiences with Hi-Fog installations, about cooperation with Marioff and about the functionality of the system are also referred to in the advertising and in communications in general.

In sum, at the launch stage, Marioff tried to gain market acceptance mainly through *building up market awareness*, during which process the *relative advantages* of the new system were highlighted through comparison with conventional systems. Furthermore, the *benefits* of Hi-Fog were made *observable* in the demonstrations. Thus, the firm had a direct influence on market acceptance.

Improving the Hi-Fog system

The prototype that was launched at the Cruise and Ferry exhibition proved too heavy and therefore was not installed. The sprinkler was re-engineered and reduced in size, and it came out in 1991, although it was officially launched only in 1992. It was the first sprinkler to be used in installations and was considerably smaller than its forerunner. Since then new and improved generations of sprinklers have come out regularly. The constant tendency has been toward smaller and lighter sprinklers that provide even greater performance and higher reliability.[21] Thus, the launch of the Hi-Fog system also seems to have characteristics of the probe-and-learn process (cf. Lynn *et al.* 1996). Moreover, the fact that the system was launched very openly before it was modified suggested *features of customer-related proactiveness*.

The high-pressure pump unit is an important part of the system, alongside the sprinklers. Over the years, the size of the unit has decreased, its efficiency has increased and its application areas have diversified. The first pump from 1991 was electrically driven, and this was followed by the introduction of a diesel pump unit and later by a self-contained gas-driven unit. A machinery space accumulator unit was introduced in 1993 and was used for small-space machinery applications and gas-turbine enclosures (see Figure 5.18).

Although at first the focus was on developing a lighter extinction system, it was soon noticed that the new system was also a very efficient extinguisher. The early success of Hi-Fog as a "different sprinkler technology" led to the rapid development of other systems for a wide range of applications. Marioff started fire tests on "halon replacement" systems for use in vessel machinery spaces as early as 1991. The system seemed to fit particularly well in machinery spaces, commercial cooking and frying systems, special hazard areas and turbines and even in engine test cells. These were all fire-protection areas, which traditional water-spray systems could not handle very successfully. The company installed the first machinery-space systems in 1992. The same year, the first order for a land-based application came in from London Underground for self-contained storeroom protection. The *emergence of Hi-Fog applications* is illustrated in more detail in Figure 5.19.

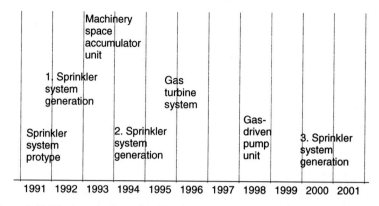

Figure 5.18 The introduction of major Hi-Fog systems and components (1991–2001).

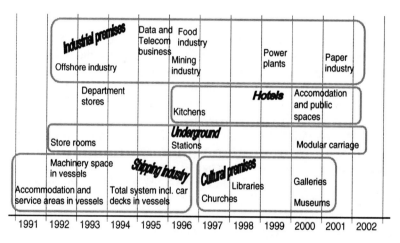

Figure 5.19 The emergence of new application areas for Hi-Fog systems (first deliveries based on Hi-Fog reference lists). This figure is by no means complete, but illustrates a broad array of different application areas.

During the development process, the system has been tested mostly to verify its suppression effectiveness in different fire scenarios. Since 1991, over 5000 full-scale fire tests have been conducted in independent testing laboratories (e.g. VTT) and witnessed by a variety of authorities (e.g. insurance companies). Cooperation with SP and VTT still continues, although Marioff's own fire-test facility, where most of tests are carried out, was established in 2001. Having its own facility enables the company to engage in more experimentation and to better maintain secrecy for the development work. It also reflects the important status of R&D in the firm.

Involvement of clients and authorities in the system development has been limited. Only occasionally, when Marioff has been developing special new systems for a particular application, have they shown more eagerness to be involved, even to the extent that it has almost disturbed the development work. All in all, the *research and development* has become very *secret and confidential* due to the negative consequences of having open doors in the early tests.

Nevertheless, when the new system is ready and already protected by patents, there is often close cooperation with the customer and joint planning. For the further improvement of the system, it has been an advantage that distribution has often been directly through client groups rather than through conventional distribution channels. Thus, the product developers get ideas and requirements for system modifications from the field, or more specifically, field workers state the problems usually faced during the installation, and the developers try to find the solutions. Modifications are realized if there seems to be market potential, preferably an existing order.

Creating the rules of the game

However, concurrently with the development work, there was another big problem to be solved. The system did not fit any standards nor did it fulfill any existing requirements because it differed significantly from existing systems. *Approval* for the first installations was needed from the SMA (Swedish Maritime Administration), and since there was no corresponding system, the approval required a *new regulation*. This process normally takes around ten years, but by *lobbying* Marioff managed to push it through in half a year. The extinction demonstrations played an important part in the process. Furthermore, the Swedish client company was also rather active in convincing local authorities. The first approval was crucial because it paved the way for coming orders.

Since regulations and standards play such a major role in the fire-protection business, it was of paramount importance to get international recognition of the Hi-Fog system. Although the Swedish Maritime Administration approved the first installations, it was necessary to obtain IMO acceptance before moving into the international field of shipping. The lack of compliance with any existing standards meant that *new standards for water-mist extinction systems* had to be set. "It's been a continuing problem. That's the problem when you're a pioneer, that you don't fit into any existing parameters and everything starts from scratch" (Manager, Research and Development). Marioff became involved in the standard-setting process in 1991 when its representatives attended the IMO Fire Protection Subcommittee meeting that was charged with developing regulations for "alternative sprinkler systems" and "alternative arrangements for halons" for passenger ships.

Because Marioff had experience with full-scale fire testing and an understanding of water-mist systems, it was able to actively take part in the process of developing the standards. Since then the company's representatives have belonged to the IMO subcommittee and have actively influenced standards in the industry. At the time, the most important IMO resolutions were carried in the years 1994–1995: guidelines for the use of Hi-Fog systems in machinery spaces were adopted in 1994, and the following year, they were extended to cover occupied areas of passenger ships.

Marioff was also active in other areas. In 1992, it approached the NFPA[22] (The National Fire Protection Association) and requested them to set up a new committee to draw up a standard for the emerging water-mist extinction systems. The committee was set up in 1993, and the standard emerged in 1996. Since then, Marioff has had a representative on an NFPA subcommittee charged with updating the standard, and it has recently also joined the water-mist task group of the European Committee for Standardization, the aim of which is to develop a European standard for a wide range of applications.

Land-based standards concerning water mist are based on IMO marine standards. Resistance has been particularly strong in the area of land-based applications: unlike in the marine business, the national legislation here is heterogeneous. Furthermore, it has been surprisingly difficult to convince local insurance companies of the advantages of the new system, and Marioff faced particularly strong resistance from Finnish insurance firms.

> Well, they'd tried to be terribly wise in their ignorance and that way put a lot of barriers in the way of progress. They'd been to England once to take a look at some fire and thought they were experts after that. [deep sigh]
>
> (CEO)

As the R&D Manager noted, this resistance may have stemmed from the fear of taking responsibility for novel, emerging phenomena. Consequently, it was only after the American insurance corporation Factory Mutual Research Corporation approved the Hi-Fog system in 2001 that Finnish insurance companies started to accept it. The VdS[23] approval granted in 2002 has considerably increased the sales of land-based applications.

Active participation in standard creation could be seen as a way of *removing the customer constraints that prevent purchasing* and thus of directly influencing demand. Furthermore, it has proved to be rather efficient in *breaking down the resistance* prevalent in the fire-fighting industry. Besides, even though being a pioneer has not been easy, Marioff has been able *to strongly influence the rules* that are now governing the industry. It has also utilized the regulations in its advertising by emphasizing the coming regulatory changes concerning sprinkler systems.

When Hi-Fog entered the market, *competitors* selling conventional sprinklers to vessels did not react immediately: the marine business was of minor significance to them since sprinklers had not been required in vessels before. "They didn't lose anything. They just didn't get on board. If they'd actually lost something, they'd probably have been more aggressive" (CEO). On the other hand, the onshore business defended its market aggressively and lobbied against changing the standards to make them applicable to water-mist systems. Firms utilizing traditional technology did not want new players to invade the market and therefore put up strong resistance to Hi-Fog.

Most competitors still concentrate on conventional water sprinklers and consider water mist as only the by-product. The first competitors in the water-mist business emerged in the mid-1990s. Marioff has welcomed the competition since it has the technological lead and extensive patent protection for its systems, and it realizes that it can even help in promoting water mist as the mainstream product in fire protection.[24]

Aiming for continuous growth

Exports have played a role at Marioff since the beginning, which is natural since the marine industry is inherently global, and today accounts for around 90 percent of the total sales. The top markets are Scandinavia, Germany, France and Italy, as well as the United States. Marioff aims to be close to its customers and prefers working through wholly owned subsidiaries. In 1995, it established its first subsidiary abroad, and since then it has set up almost 20 subsidiaries around the world. These subsidiaries mainly provide technical support and assembly services and take care of the growing after-sales service. Head office concentrates on sales to the shipping industry, while the subsidiaries take care of most of the onshore sales. This reflects the substantial difference between these two markets and the consequent need to adapt land-based applications to the various specific standards.

Marioff also has distributors and agents worldwide. According to the R&D manager and the CEO, it is important to have local employees who know how best to act in the particular market and who are able to develop close contacts with local authorities. In the future, the company aims to further increase sales through organic growth and geographic expansion, and in this it has been an advantage that sales are not dependent on any certain region or business.

Hi-Fog currently has over 90 percent of the market for cruise- and passenger-vessel sprinkler systems and is increasing its hold in land-based applications. Getting a steady foothold in the land-based market is important, since the total volume of the marine market is limited. According to the CEO, Marioff will continue to keep its focus on water-mist fire-protection systems. He also believes that future prospects for the Hi-Fog system are bright, since environmental and safety concerns, and the legislation, are likely to become stronger, which will thus increase the demand for water-mist-based extinction systems.

5.4.7 Customer-related proactiveness during the Hi-Fog development process

In sum, the development of Hi-Fog from an idea to an established system took only four years. In fact, the time between the idea and the ready system was only four months, the rest of the time being spent on establishing the system in the market. The time line is shown in Figure 5.20.

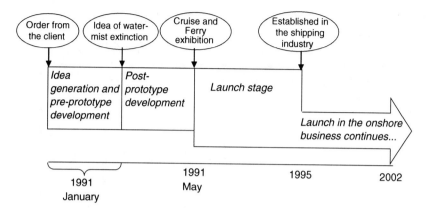

Figure 5.20 The innovation development stages of the Hi-Fog system.

The process of developing the Hi-Fog system seems to reflect different degrees of customer-related proactiveness (see Figure 5.21). At the idea-generation stage, the firm was searching for an answer to the client's problem. Since the client's request was undefined and the idea for the new extinction system was generated inside the firm, we can conclude that the firm was *anticipating* the market opportunity at that stage.

Since the firm was trying to solve the client's problem by trying out different extinction systems, the idea and the development of the prototype were almost simultaneous, and thus ideation and pre-prototype development could be seen as one phase. Once the prototype had been constructed, the firm *reacted* to the needs of the customer and developed the system even further.

Customer-related proactiveness increased at the launch stage, and the firm actively *influenced* demand by both creating awareness about the new system and removing customer constraints preventing purchasing, i.e. influencing regulations and standards. It was also constantly modifying the system after the launch, which indicates that customer-related *reactiveness* was also high at the launch stage.

Although there was only one client at the early stages, the orientation of Marioff seems to have been *geocentric* throughout, reflecting the inherent global nature of the shipping industry. Consequently, it seems appropriate to slightly modify the a priori framework by including *the characteristics of the industry.*

Hi-Fog was created at the time of *environmental turbulence* in the fire-protection industry. This turbulence was due to the two substantial changes, the growing importance of environmental concerns and the tightening of safety regulations, both of which contributed to the need for a new kind of extinction system. Turbulence was further accentuated, since traditional players in the fire-protection industry seemed to largely ignore the coming changes.

There were many organizational variables in Marioff that seemed to contribute to its customer-related proactiveness in particular. Its organizational

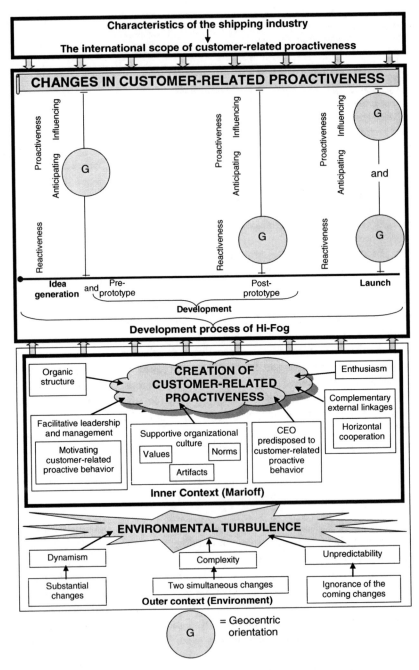

Figure 5.21 Customer-related proactiveness during the development process of Hi-Fog.

structure was apparently rather organic, and proactiveness was also evident in its values and norms, even in its artifacts, since there were stories reflecting this kind of behavior. The CEO himself seems to be a proactive customer-oriented person, and this also helped. Furthermore, his *enthusiasm* for the new system that was apparent during the ideation phase seems to have fostered an eagerness to build up awareness about it when it was launched. Enthusiasm is thus added to the framework.

However, the firm missed some factors that, according to the theory, tend to increase proactiveness. First, it did not emphasize the building up and use of customer-related proactive capabilities. Second, complementary external linkages were not particularly accentuated in this case. Third, the decision-making in the firm has been rather autocratic, although this autocratic decision-maker, i.e. the CEO, has a very proactive personality and is a very efficient motivator of proactive behavior throughout his firm. Fourth, although Marioff had enough liquid financial assets, it did not seem to focus on increasing proactiveness toward its customers.

5.5 Nordic Walkers

5.5.1 Nordic Walkers as a radical innovation

People have been using walking sticks from time immemorial, their basic function being to facilitate moving in rough areas. In the early 1900s, a small number of cross-country skiers began to use their ski poles on hilly terrain during their summer training. On a minor scale, poles have also been utilized in some strenuous exercises requiring maximum physical exertion, such as power walking,[25] which is fast walking carrying added weights and with the arms at a 90-degree angle. However, the idea of *creating special poles that everyone could use throughout the year for exercise purposes* was born in 1997 in Finland. The Finnish company Exel plc. was the first to develop and launch the Nordic Walker, a pole specifically designed for fitness walking (see Figure 5.22).

Nordic Walkers gained *commercial success* extremely quickly. The product was launched in Finland in 1997, and by 2004 around 1.5 million Finns had at least tried the sport during the previous year, and almost 0.8 million were practicing it regularly (see Figure 5.23). Although growth is now slowing down in Finland, Nordic walking is increasing strongly in other countries, particularly in Scandinavia and in German-speaking parts of Europe.

The *technological novelty* of Nordic Walkers stemmed from the need to design a new kind of pole for a new sporting activity. New poles were needed because, first of all, the walking technique differs from the skiing technique. Second, Nordic walking can be practiced in very many places (e.g. woods, sawdust tracks, sandy beaches, city streets, mountains), whereas skiing tracks do not differ that much, and third, it is not dependent on the season. Consequently, Nordic Walkers differ from skiing poles in material and form. They are usually made of composite material, carbon fiber or glass fiber, in other words material

Figure 5.22 The Nordic Walkers product.

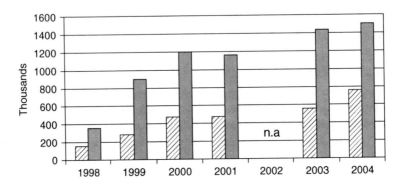

☑ Regular practitioners, i.e. those practicing at least once a week

▣ Experimenters, i.e. those who had tried it at least once during the previous year

Figure 5.23 The adoption of Nordic Walking in Finland, 1998–2003 (The Central Association Internet pages; Nummela 2003; Jantunen, telephone discussion February 16, 2004).

that is durable, light and does not vibrate when the pole strikes the ground. The forward-angled lock spike tip touches the ground firmly and at an angle that suits the walking technique, while Asphalt Paws are used for street walking. Furthermore, the grip and the strap are ergonomically designed to enable a relaxed grasp that does not make the shoulder muscles sore.

The utilization of Nordic Walkers, in other words Nordic walking, incorporates certain *substantial benefits* for the adopters. It seems to provide a new kind of flexibility for training purposes, and unlike many sports, it can be practiced almost everywhere. People are able to start from their front door. It enables social intercourse, i.e. talking while taking exercise. Furthermore, by adopting different techniques, people in different physical conditions can walk together and can all benefit from it. Thus, it suits people in very different states of fitness and health. Studies (e.g. Church *et al.* 2002) also indicate that people do not feel that the exercise is as hard as it really is. Hence, Nordic walking produces more energy output for the same perceived level of exertion.

The launch of Nordic Walkers has also *changed the behavior* of the adopters. It created a completely new sport that is practiced so extensively that it has even changed street scenes in countries in which it has gained popularity. It has also encouraged a whole new group of people to engage in fitness training.

Nordic Walkers have certain special features that also seem to have influenced their development and marketing. First, the product is a consumer durable that normally lasts several years. Compared to other sports equipment, its price is affordable (25–100 euros). Moreover, as a sports device, it has two dimensions: on the one hand it is bought for leisure activities, and on the other hand it is acquired with a view to improving one's physical condition and, consequently, one's health. Unlike many other types of sports equipment, it is appropriate for a very heterogeneous group of people, i.e. the masses.

5.5.2 Exel plc. – the developer of Nordic Walkers

Exel plc. designs, manufactures and markets composite products (carbon and glass-fiber-reinforced plastics) for sports, leisure and industrial applications. The firm was founded by three chemists in 1960 and specialized in the production of electronic detonator caps: the name Exel stems from the words "Explosive Electronics." Product development took it into sports, and in 1973 it started to produce cross-country ski poles. The manufacture of industrial applications started in 1980, and nowadays Exel is the largest pultruder in Europe.

The company strategy is based on specialization and it aims to be the market leader in its chosen segments that use composite technology. Exel was listed on the Helsinki Stock Exchange in October 1998, and its largest shareholders in 2006 were the Swedish mutual fund Nordstjernan AB (29.4 percent), and the Finnish mutual pension insurance companies Ilmarinen (6.8 percent) and Varma (4.3 percent). At the turn of the millennium, the firm consisted of the Sport and Industry Divisions, both of equal strength. As this study is concerned with walking poles, the focus is on the Sport Division (see Figure 5.24), leaving the Industry Division out of the picture. Thus, the word "Exel" henceforth refers to the Sport Division unless otherwise specified. The product applications of the Sport Division are poles (cross-country, alpine, trekking, Nordic walking), water sports (windsurfing masts), ski-, snowboard- and ice-hockey stick laminates, floorball, Finnish baseball, composite hockey shafts and floorball clubs with related accessories.

The organizational structure in Exel seems to have been rather *organic*. Inter-firm communication has been open, and the information has flowed continuously between the marketing and R&D functions. Even though, in general, people working on product development may not have been as extrovert as those

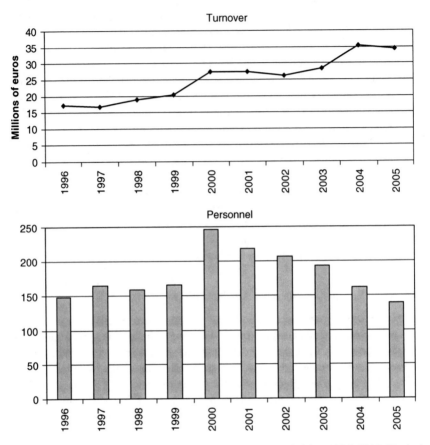

Figure 5.24 Turnover and personnel of the Exel Sports Division 1996–2005 (Exel plc. Annual Reports 1997–2005; Huoso, e-mail September 9, 2003).

working in marketing, the joint interest in sports has often paved the way for interaction and cooperation. The emphasis has been on getting things done rather than on formal bureaucracy. Subordinates have been involved in the decision-making, and in their annual individual appraisal discussions, they tend to bring up rather openly ideas for altering the organization or changing the procedures. In cases of failure, it has relied on open communication and learning from its mistakes. Blame is rarely apportioned, and employee turnover has been low.

Sports-minded applicants have been favored in the recruitment policy, and many employees are former athletes: physical education has been seen as an advantage.

I think it's important that if a company manufactures sports equipment, that its people also have to be sporty and outgoing and multilingual. I don't think it's necessarily very good for your credibility if someone who is 30–40 kilos overweight and sweating heavily, tries to sell you something relating to sport or exercising at some trade fair.

(Senior vice president, Exel)

Favoring sporty job applicants could be seen as an indication of customer orientation in general rather than of customer-related proactiveness in particular. Staff training in the company seems to be well organized and is based on requests from employees; consequently, language skills and commercial training have dominated. Training based on employee requests is not likely to emphasize proactive capabilities, since it is highly unlikely an employee would admit that he or she needs training in problem finding or creative thinking. Thus, *building up customer-related proactive capabilities* has not been particularly emphasized in the selection and training of employees. However, customer-related proactiveness was facilitated through fostering inter-functional cooperation. The aim of the management has been to generate growth through new innovative products, and this is communicated throughout the firm. Furthermore, the importance of knowing the customers' behavior and latent needs was acknowledged in the innovation process. In general, *enabling proactiveness* was demonstrated in the resource allocation and in the top management's encouragement to go ahead with the product development.

Exel uses incentive payments to reward its employees for successful work. However, it became apparent in the interviews that although monetary rewards are important, they are not always the main motivator. Motivation also seems to arise from being able to combine one's work and hobby, from being able to work in the sports business with other enthusiasts and from the feeling that one is able to raise the quality of life of customers by getting them to exercise more.

A lot of people have really put their hearts into this, because you're basically working for a good cause. Even though the commercial side of it is always in the background, the main driving force is that we're selling equipment to make people feel good with our products.

(Senior vice president, Exel)

Thus, the motivation in Exel seems to be connected to customer-related proactiveness.

Open internal communication and high employee responsibility have been *valued*, which indicates that the premises for customer-related proactiveness existed. The continuous development of new products has been regarded as an important source of competitive advantage, and discovering latent customer needs has been a crucial part of the planning process. Since competition in the sports-equipment market is hard, it has been rather clear that the firm cannot limit its efforts to expressed needs. This also seems to have been reflected in the *norms*, and customer focus has been prevalent. The interviews featured plenty of stories and anecdotes about influencing the market, building up demand and teaching the market to use new sports devices, and all of these *artifacts* seem to indicate customer-related proactiveness.

In general, Exel employees were considered open, active, creative and willing to do their best to get the customers to exercise more and to do it with products that suit their particular needs best. Thus, they seem to have been *predisposed to customer-related proactive behavior*. This leads to an interesting question: if proactiveness is not especially emphasized in the employee selection or training, how do these proactive individuals end up working in the firm? The answer may lie in the firm's strong emphasis on hiring sporty people. It seems that the general characteristics of sporty people, at least, partly coincide with those of individuals behaving proactively toward customers (see Section 3.4.4). Past studies[26] indicate that sports-oriented participants tend to be more achievement-motivated (Reiss *et al.* 2001: 1143) and more extraverted (Courneya and Hellsten 1998: 630; Ingledew *et al.* 2004), and they are also likely to have a higher need for activity than non-exercisers (Rhodes *et al.* 2004). Moreover, athletes are also likely to be conscientious, since they are used to setting high goals and doing their best to achieve them (Jones 1996).

This section briefly described organizational factors related to Exel. The firm's environment at the time when the development of Nordic Walkers began is considered next.

5.5.3 Environmental turbulence

Exel ski poles came into prominence in the Innsbruck Winter Olympics of 1976, when 40 of the 56 medal winners used them. Thus, the brand quickly became well known among cross-country skiers throughout the world. Market share grew steadily during the 1980s, but in the 1990s, the overall market for cross-country skiing equipment began slowly to diminish. Although Exel's share of the world market was still growing, the *market situation was becoming tighter* and *price competition tougher*. The competition for market share in sports equipment has generally been fierce, and a few big global players such as Nike and Reebok have dominated. There is, however, a heterogeneous group of smaller companies operating in various niche markets in which the big players have no interest.

The decline in cross-country skiing as a hobby happened for various reasons. For one thing, there were many successive winters with very little snow in traditional skiing countries in Northern Europe and North America. Second, many new sports (e.g. downhill skiing and snowboarding) emerged and provided new alternatives for the sports-minded. Furthermore, the image of cross-country skiing was becoming somewhat outmoded and unfashionable, especially compared with the images of the new sports. Thus, the general trend toward less interest in cross-country skiing was evident. At the same time, the growing amount of leisure time and the increasing numbers of wealthy, hale and hearty senior citizens created new potential target groups for sports-equipment manufacturers.

Thus, sports-equipment firms encountered big but low-frequency changes in their environment as new sports emerged and the demand for old ones slowly decreased. Similarly, the increasing amount of leisure time and the growing numbers of active wealthy retirees could be seen in terms of substantial but slow changes. Thus, the *dynamism* was somewhat high. Since the above-mentioned changes were often interrelated (e.g. increasing wealth facilitates the practice of new sports), the *complexity* was also high (see Figure 5.25).

Most of these changes were well known and predictable, which seems to have decreased the degree of environmental turbulence. Nevertheless, in the case of Exel, in which the demand for sports equipment was largely dependent on the snowfall, the unpredictability of this particular factor contributed to the overall environmental turbulence.

5.5.4 Idea generation

Since the market for cross-country skiing poles was gradually diminishing and the competition in the sports-equipment market in general was getting harder, it was clear to Exel that additional growth should be sought in new products. The search for new ideas was rather systematic in the middle of the 1990s. Trends, such as the increasing numbers of elderly people, global warming and the increasing popularity of walking as a hobby around the world, especially in the United States and Japan, were analyzed, and attention was consequently directed toward walking. The challenge was to combine this with Exel's traditional know-how (pole manufacturing).

Exel was not alone in developing ideas for new sports: two other Finnish organizations (The Central Association for Recreational Sports and Outdoor Activities, and the Sports Institute of Finland) were also actively working toward this end.

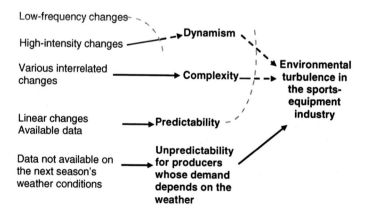

Figure 5.25 Factors contributing to the environmental turbulence in the sports-equipment industry (the degree of unity of the arrows illustrates the degree of influence of the factors).

The Central Association for Recreational Sports and Outdoor Activities (Suomen Latu ry) is a non-profit organization that aims at encouraging people's interest in outdoor activities and sports by arranging and supporting outdoor sports and health-promoting events. The Association consists of 230 member organizations in various parts of Finland and has about 70,000 members. It made some attempts at promoting walking using skiing poles in the late 1980s, but interest remained low. One of the first events took place in January 1988 in Helsinki:

> We built a statue for Tahko Pihkala next to the Olympic Stadium and we decided that everyone should come to the inauguration with a piece of sports equipment. Then we put a notice in the newspaper that we'd start off at Maunula cottage and ski on cross-country skis from there to the venue. But then the snow melted … and that was in January! In the end we decided we'd do it anyway – if we'd said we'd start off at Maunula, that's where we'd start off. Then I managed to get hold of [ski] poles and everyone just walked from there. I think we were about 50. And you could say that it was the first event, at least as far as I know.
>
> (Secretary general, The Central Association)

This event was filmed by Finnish television, walking with poles was regarded by the public as a curiosity, and practiced by people who "left their skis at home." Nevertheless, a few people in the Association believed in its potential, especially since winters without snow seemed to follow one after another:

> Then we realized that there were all these skiing trips and skiing events and they couldn't be organized. When the snow melted, it had a horrible effect. After all, we wanted to organize events, with or without the snow. The idea wasn't originally that we'd develop a good form of exercise for the average person.
>
> (Secretary general, The Central Association)

Thus, the first interest in walking with poles in the Association stemmed from its *big potential*, i.e. its applicability to all kinds of weather conditions. At the same time, various Finnish Sports Institutes were using poles when they trained athletes. The Sports Institute of Finland was particularly influential in creating special poles for walking. Situated in Vierumäki, it is a centre of education, leisure and sports that develops, produces and markets training, exercise and educational services in the fields of physical education and leisure. The institute has more than 300,000 visitors a year, and besides training coaches and athletes, it is responsible for the physical training and preemptive health care of the personnel of various firms.

It was common to see cross-country skiers practicing with skiing poles around the institute during the summer season, usually when they were running up hills. However, the head of physiological testing was not interested in training top athletes but rather wanted to encourage the average person to take care of his or her health: he was on the look-out for new training methods that would *suit ordinary people*:

> I was thinking about types of exercise you could offer ordinary people as a training method and then I came across a survey where walking was number one. I started thinking that hey, walking – everyone knows how to do that, don't they? But how could you improve and add value to walking? Then I saw a group of cross-country skiers walk past me with skiing poles and thought [claps hands together], oh my God, would Finns be crazy enough to use poles?
>
> (Head of physiological testing, The Sports Institute of Finland)

He saw the benefits of walking with poles to be especially in the exercising of various muscles in the legs and the upper body. He was so convinced by this that at the beginning of the 1990s, he ordered from Exel 20 or 30 pairs of poles that were ordinary skiing poles, but shorter. They were used by groups of ordinary people who came to the Institute on fitness-rebuilding programs, and the feedback was rather encouraging: "The whole time the feedback was 'bloody excellent! ... But you wouldn't want to do this in broad daylight, it looks awful!'" (Head of physiological testing, The Sports Institute of Finland). Although at that time there was no hint of the coming big success of walking with poles, he was convinced that this new sport would be particularly suitable for ordinary people.

> In my line of work I can see that Finland is split 20/80 – 20% are those who were satisfied with the services the leisure-sports industry was offering: step aerobics, spinning, gyms ... But the so-called average Joes, those who don't like that kind of sport. There were two million of them and all of them walking! So then I thought, that's where we want to be! That there was a clear demand for a pleasant experience from doing exercise, something these people had never been offered.
>
> (Head of physiological testing, The Sports Institute of Finland)

The tests that the head of physiological testing conducted with these groups of people clearly indicated that walking with poles was very effective in raising the pulse to the level that benefits the heart and the circulation system. Encouraged by these positive results, he telephoned Exel's testing manager and urged him to develop a special kind of pole suitable for walking. The idea was not new to Exel, since they had been looking for new products that would exploit the company's know-how (e.g. pole manufacturing). As a result of their trend analysis, they knew that walking was gaining in popularity. They had also found out that Nordic-Trek-type exercise machines were becoming more and more popular in gyms, and the exercise involved was somewhat similar to skiing and walking. Nevertheless, the encouragement from the head of physiological testing was important in convincing Exel that there might really be a demand for this kind of product.

At the same time as Exel was deliberating the potential of walking poles in the spring of 1997, the testing manager received a call from the secretary general of The Central Association for Recreational Sports and Outdoor Activities. The Association was planning to include in its own magazine *Latu ja Polku*, a brochure demonstrating the benefits of walking with skiing poles. They wanted to take the necessary pictures in the Sports Institute and asked Exel to provide them with the skiing poles and the Institute to provide the athletes to be used as models.

Consequently, in early April 1997, Exel's testing manager, the Central Association's secretary general and the Sports Institute's head of physiological testing met at the Sports Institute and followed the photographing. The athletes were performing uphill–downhill exercises taking long strides, which was still reminiscent of traditional summer training among cross-country skiers. After the session, these three key people started to discuss how to go on with the development of the sport, and they agreed that the ordinary skiing pole was not optimal for walking. Exel's testing manager then offered to make a special pole for the new sport and started the development work as soon as he arrived back at the factory. Thus, this photographing event was significant in that, besides producing a brochure, it also resulted in an idea for new sports equipment and strengthened the motivation in these three organizations to jointly promote walking with poles.

In sum, the idea of creating special walking poles for the masses was born in early April 1997. It can be said that market opportunity was mostly *anticipated*. Certain people in Exel, the Sports Institute and the Central Association had a presentiment that this new sport might become a success, a feeling that originated from their previous experience with sports trainers. The anticipation was also based on the systematic search for new sports opportunities in Exel. Given the fact that information on global trends and circumstances in different markets was extensively used, the orientation of the firm could be considered *geocentric*.

5.5.5 Development

The developmental challenge was that it was not only the product but also the sport that had to be created. At first, the market research did not seem to be appropriate in terms of anticipating demand because not even the sport existed.

The feedback that the head of physiological testing at The Sports Institute of Finland had collected from groups walking using ordinary skiing poles was beneficial in developing the prototype of the walking pole: the strap on ordinary skiing poles seemed to cause muscular pains, since it forced the walker to press on the pole, for example. Thus, the cooperation with the head of physiological testing was valuable. Moreover, since Exel had solid experience with skiing poles, it was able to use this to support the design of the special walking poles. The first prototype was developed very quickly and was ready in June 1997.

Walking-pole prototypes were first used and tested by the people who participated in the activities organized by The Sports Institute of Finland. According to both Exel's and the Institute's representatives, the cooperation was trouble-free, and information was shared openly. Those who tried the sport gave constructive feedback, which was carefully taken into account in the prototype modification.

> They gave us really good advice, as in "do this and this, I noticed this and that thing doesn't really work". So, really, it was the users, they were a really good help in developing this kind of product. You could really say that 80% of the product development was actually done by users.
>
> (Head of physiological testing, The Sports Institute of Finland)

The eagerness of the first users to give feedback and to participate in the product improvement indicates that they were also rather enthusiastic about the new product. Requests for modification were mostly useful and led to the development of adjustable straps and the removal of the baskets from the poles, for instance. The development of the sport was concurrent with the development of the product, which meant that the firm was given valuable information regarding the potential of both during the tests. The feedback concerning the new sport was very encouraging, and its suitability for ordinary people was further assured. Nevertheless, it also became evident that the biggest challenge would be to get people to walk with the poles in public. "Well, it really was quite awful … For example, that people were heckling when we passed the golf course, they were going 'hey, have you forgotten your skis?'" (Head of physiological testing, The Sports Institute of Finland). To conclude, throughout the development stage the developers seem to have *reacted* to the needs of the end customers. Reacting was, to some extent, possible before the first prototype development, since the firm was able to use feedback from people walking with skiing poles as a starting point. Although, as mentioned earlier, the ideas for the new product were sought and weighted based on global information, feedback at the development stage was collected only from Finnish people: thus the orientation was *ethnocentric*.

5.5.6 Creating the market

The launch is regarded to have begun in autumn 1997 when The Central Association for Recreational Sports and Outdoor Activities started to organize Pole Walking nights. At the same time, the brochure that was photographed in April at the Sports Institute came out as an appendix of the *Latu ja Polku* magazine, in which the first advertisement for walking poles also appeared. In general, the launch was a demanding task, since the strategy that was commonly used in Exel was not suitable for this new invention:

> In a way this was a special project for us. Usually companies manufacture products to fulfill a certain demand and then market the product, talk about its technical properties. But in our case, we actually had to develop a sport, market it and invent a product for it.
>
> (Senior vice president, Exel)

Thus, at first it was necessary to get the people interested in the new sport.

Enhancing interest in walking with poles

Exel launched its walking poles under the trademark "Walker," and the first production run was only a couple of thousand pairs. At first it seemed difficult to get even that amount sold, since the retail trade did not believe in the product and was not keen to take it onto their shelves.

> In the beginning we made a … well, not necessarily a mistake, but … We introduced this to Finnish store executives by saying that walking with poles could become something big, as in how about it if we start taking this further together. They practically laughed in our faces, they thought nobody would start walking with poles!
>
> (Senior vice president, Exel)

Therefore, it was necessary to build the demand by focusing directly on the consumers and *making them want to adopt a new sport*. This desire was created through building awareness and giving people a chance to try it out. It was a particularly demanding task, since the first target group consisted of people who were not usually enthusiastic about new sports:

> For the most part we've been trying to reach people who are, what you might call non-active, couch-potatoes, people who play with the remote control. There's an amazing potential there, and it's the biggest target group in the world at the moment.
>
> (Senior vice president, Exel)

Motivating these people was not thought to be an insuperable problem, since walking with poles could be shown to be closely connected to health.

> The papers are increasingly writing stories about people with high cholesterol, heart disease and the like. These people need to be reached through the media and we should try to get them to understand that here's a really easy way to gain control of your own body and get fit.
>
> (Senior vice president, Exel)

Awareness of the sport was mainly built up through mass events and advertisements. Mass events, such as Pole Walking Nights, were mainly organized by The Central Association for Recreational Sports and Outdoor Activities, and the first ones proved to be a success:

> We placed an ad in *Helsingin Sanomat* in early August and tried to lure them in with juice and coffee and in any way possible, really. And then we were really amazed how many people actually turned up at an event such as

this. And even the next week there were so many people that we had to stop serving juice – it was just impossible to continue to do that with such a big crowd.

(Secretary general, The Central Association)

These events have been organized regularly in various places around Finland since August 1997 and have been particularly popular among middle-aged women. Exel provides their walking poles for loan on these occasions.

Some print advertising in special-audience magazines was also used, which enabled the targeted message to reach people with a particular interest in outdoor activities, rheumatics or cardiovascular diseases, for example. Thus, marketing concentrated strongly on the health benefits of the new sport, and the technical details of the poles played a minor role. Cooperation with health associations (e.g. The Rheumatism Association, The Allergy and Asthma Federation) was utilized to convey the message that walking with poles was good for the health. To further enhance this message, Exel actively used the results of international scientific research proving, for example, that the poles induce the walker to use more oxygen and burn more calories than regular walkers traveling at the same time. Support also came from various doctors who publicly highlighted the benefits of the sport for the health. The interviewees described the utilization of doctors in the advertisements as follows: "It gives you credibility. People really do believe men and women in white coats" (Senior vice president, Exel). "In terms of marketing the sport, I think it's been great. It gives you a ... how should I put it ... that the nonsense disappears from around it. People are trustful" (Head of physiological testing, The Sports Institute of Finland). The Internet was also utilized in communicating the benefits of the new sport. It is worth emphasizing that, regardless of the media, in the early stages the message concentrated on the new sport, and the equipment played only a minor role. There was also a massive amount of "free" publicity in the media, which was keen to cover "the new strange sport" and to show people practicing it:

We got more and more excited because the media was with us, that we didn't need to try to get them interested. All we needed to do was to tell them about this amazing thing, and all radio networks, television stations ... all came after one phone call.

(Secretary general, The Central Association)

Although the general media were interested, the sports editors were not so sure at first: "They thought this was something completely mad. Not even a sport ... Well, maybe it was good that it all started as something else than a sport" (Secretary General, The Central Association). In sum, various ways of promoting interest in Nordic Walking were utilized simultaneously. Nevertheless, awareness was not usually enough to get people to buy the poles: they also needed to be taught to use them.

Teaching people to walk with poles

Thus, the biggest problem was not awareness but *getting people to walk with poles*. This was particularly challenging since the sport was meant to be an outdoor pursuit, along trails and on pavements, and people often fear looking strange in public. Furthermore, in order to get people to buy the poles, it was necessary to get them first to try the sport. The mass events organized by The Central Association for Recreational Sports and Outdoor Activities were also important in this respect. This was where people had the opportunity to try the sport, and they were taught to use the poles properly.

In order to reach even larger numbers of people, it was decided to build a sport-instructor network. The cooperation of the sports and health associations was sought in finding instructors from all over the country, in educating them about the new sport and in providing them with training material and poles. The role of The Central Association for Recreational Sports and Outdoor Activities was pivotal in building the instructor network and producing the training material. The network was built quickly,[27] since people were enthusiastic about the sport and thus keen to become instructors. They then lent poles to people and taught groups how to use them properly: although the walking was not difficult, people had to learn the right technique. The Sports Institute of Finland continued to introduce the sport to approximately 100,000 visitors a year. The instructors had an important role in giving people a positive experience of the sport. When they tried it properly, they noticed the benefits:

> After they'd finished exercising, they realized how little effort was needed in it, even though it still made them sweat, got them out of breath and, most of all, made them feel good ... We started getting comments from people that "hey, this is really something!" It's just a matter of getting it into people's minds that, even though it looks silly, the effects are twenty to forty percent bigger than in walking, and in the same amount of time, and that they'd be stupid if they didn't go for it. Well, obviously I didn't say it exactly like that, but it was the sort of way we sold it to them.
>
> (Head of physiological testing, The Sports Institute of Finland)

At first, the sport was mainly practiced in bigger groups and in the woods, because it was clearly easier to "look strange" in groups than alone. Later, smaller groups or pairs walking with poles emerged on the pavements, and slowly the threshold for using the poles in public became lower:

> A big group gives power to the individual, so naturally at the beginning the groups were pretty big. Mostly people would walk on the paths in the woods, away from the eyes of the world. Little by little, when people started being more aware of the health benefits of Nordic walking, the group sizes started getting smaller and people came out of the woods to walk in public places. The greatest thing was so see a man walking with poles alone in Helsinki city centre.
>
> (Senior vice president, Exel, e-mail 13.2.2004)

Furthermore, word-of-mouth communication seemed to play an important part in the innovation diffusion. Exel believed that it could also influence the message here because instructors organized sports events at which people could try the poles and get positive experiences. Thus, people tried the sport, liked it and went to buy their own poles. It was only then, as a result of accumulating direct consumer requests, that the stores were convinced to include poles in their range of goods.

> They had people coming in their store going "have you got any walking poles?" and at first, many offered them hiking poles and the like, but it worked out pretty well when people started insisting that "they have to be Exel walking poles." And then the store managers started calling us that "we've got some people here who want to buy those walking poles of yours, would you mind sending some, please?"
>
> (Senior vice president, Exel)

"It all happened really quickly – when I asked how it was going, they told me that they'd hired so and so many people, that they were selling them everywhere like crazy, that they had difficulties in delivering" (Head of physiological testing, The Sports Institute of Finland). Demand started to grow in the winter of 1997, and growth further accelerated in 1998. The big success of the new sport surprised all the parties involved in the launch. The competitors were also caught by surprise: although the first ones emerged relatively quickly, Exel seemed to have significant first-mover advantage:

> In a way, we're in a good situation if you consider other manufacturers that came after us. In practice, they've just copied our brochures, simply copy-pasted them. Then they've tried to copy the product as much as possible. [...] They needed to import the products and then try to sell them maybe a bit cheaper than we do. Whereas we have this whole network, all these people who are loyal to this brand.
>
> (Senior vice president, Exel)

Although, in principle, Exel welcomed the competition since it was also likely to help spread knowledge about the sport throughout the world, it also had its drawbacks. The main problem was the potential damage to the image of the sport caused by copies of inferior quality. This also worried the head of physiological training at The Sports Institute of Finland:

> A whole market of bad products emerged. Such bloody awful products as you could ever imagine. And dangerous products even. I mean, I don't really mind the fact that other brands come along, but you've got to have some sense of responsibility that you don't just put anything out there.
>
> (Head of physiological testing, The Sports Institute of Finland)

As the competition emerged, it became more and more important to target the sport on different practitioners.

Targeting the sport

At first, approximately 70 percent of the practitioners were women, most of them past middle age. Other age groups soon became interested, and the sport's suitability for all ages has been its advantage. In 1998, 4 percent of Finns practiced Nordic walking regularly (at least once a week), and in 2004, 19 percent were regular practitioners and 37 percent of Finns said that they had done it at least once within the past year. It is also worth emphasizing that the proportion of regular practitioners was fairly high (25 percent) among people over 65, which is not a group that is usually the first to adopt a new innovation.

> How important it is to them [elderly people] in a way, that when you wouldn't really want to use any of these walking aides that label you an old person, such as a single walking pole or a zimmer frame. Now these people have realized that Nordic walking is something for them, even though the walking poles more or less act as a balancing aid to help them walk more safely. It's still an important thing for them mentally, that they do Nordic walking just like other people, even though they walk slightly slower, but that they can still get some fresh air and go to the corner shop a lot more easily.
>
> (Senior vice president, Exel)

The suitability of the sport for ordinary people also brought with it certain problems:

> Of course, people thought of it as a sport for unfashionable people, which meant that we had a lot of work to do to overcome this image problem. We then tried to provide scientific and physiological facts and emphasize the health benefits, and prove the significance of it all to them. When you've convinced someone, they clearly don't care about what other people say.
>
> (Head of physiological testing, The Sports Institute of Finland)

Exel not only highlighted the benefits of the new sport, it also worked on the image by adapting the product for special groups. In particular, it produced small batches of specially designed poles for people with Parkinson's disease and for the blind. Although these products were not particularly lucrative in an economic sense, they were important for the image.

Over the years, the marketing has focused more on the poles and their technical features because people already know about the sport. *Different types of walking* have been developed for people with different fitness levels, and consequently, the variety of poles has increased. Special models have been designed for active and ambitious walkers and for keep-fit enthusiasts, for example. The big challenge lies in reaching these different segments with targeted messages. Traditionally, Exel's marketing has identified with world-class sports and top-ranking athletes,[28] and its advertisements have promoted the image of hard training, appealing to athletic-looking young and beautiful people. This kind of advertising is likely to get through to active and ambitious walkers but may not motivate the masses to start taking care of their health through exercising.

Going international

After the positive feedback from the home market, Exel started to plan its launch abroad. This started in 1998 and is still going on. In order to gain acceptance abroad, it called the sport "Nordic walking," and the trademark of the poles was consequently changed from "Walker" to "Nordic Walker." The new name worked well:

> Even we have been surprised by how easily the words Nordic and Nordic walking have been accepted in the countries we've been importing to. They seem to evoke the associations people have with the North, the Nordic countries – fresh air, pure nature, healthy lifestyle, etc. It's been pretty easy to get those ideas through to people.
>
> (Senior vice president, Exel)

Although the vast potential in foreign markets could be foreseen, the firm launched the product first in the Nordic Countries and only after that were more distant markets approached.

> Well, it was a pretty natural choice to us, really, because the Nordic countries ... well, we're used to skiing a lot, and even though we didn't really mention skiing when we were marketing the product ... But we thought that because Nordic walking really is a Northern sport, we thought it was easiest to start here.
>
> (Senior vice president, Exel)

Further reasons for the incremental market creation were resource constraints, the lack of competition and, especially, the need to establish a special network in each country. Although *liquid financial assets* have been enough for the innovation development, Exel admits that the launch might have been more extensive if there had been more resources to build awareness and to teach customers how to use the product in various countries simultaneously.

The Nordic markets were approached together with The Central Association for Recreational Sports and Outdoor Activities, which contacted the corresponding associations in other countries, and together they organized mass events at which Nordic walking was presented. It also educated trainers from other Nordic countries and presented the sport at various international meetings and conferences.

Switzerland emerged early as a key target country, since the local distributor was extremely enthusiastic about the sport and also capable of conveying this enthusiasm to its customers. This was then followed by Germany and Austria. Since Western and Central Europeans had been used to using trekking poles, it was considered important to highlight the fundamental differences between trekking and Nordic walking.

> Trekking is trekking, where you walk in the wild, and there the pole serves as a balancing aid and helps in carrying heavy loads in difficult terrain, whereas Nordic walking is more of a fitness-type of walking with poles specially designed for it.
>
> (Senior vice president, Exel)

The suitability of Nordic walking for people with different fitness levels was known, but applicability in different climates and different cultures surprised even Exel:

> Australia for example: when you think of it, you could never think of a reason why they'd be interested in something like Nordic walking. They don't have this kind of nature, or the kind of culture that we have here in the Nordic countries. But it did break through there as well, and people walk mostly on beaches on soggy sand.
>
> (Senior vice president, Exel)

All in all, it is estimated that in 2007, about eight million people around the world were practicing Nordic Walking regularly. The major growth has been in Western Europe, but the more distant big markets are increasing in significance. In particular, the United States and Japan are expected to be significant in the long run. However, although the markets were treated individually, their creation in each country went through the same stages. Basically, the launch strategy that was so successful in the home country was also used in other markets. Figure 5.26 describes the market-creation stages, the main means utilized and the state of launch in different countries in 2002.

As in Finland, "free" publicity for the sport was also easily gained abroad, and Exel has managed to make the most of this. In 1998, a photograph of Pope John Paul II walking with the poles was broadcast all over the world in newspapers. Other well-known proponents of the sport include King Carl XVI Gustav of Sweden, two presidents of Finland and HRH Prince Edward, the earl of Wessex. Some advertising was also used in the early stages, but it concentrated mainly on emphasizing the health benefits.

Teaching Nordic walking and building awareness of both the sport and the equipment happened in cooperation with national sports and health associations. There has been a need to find, motivate and educate instructors in each country, although the instructor networks are country-dependent: "In some cases we build the organizations simply by starting looking in various places – the teams

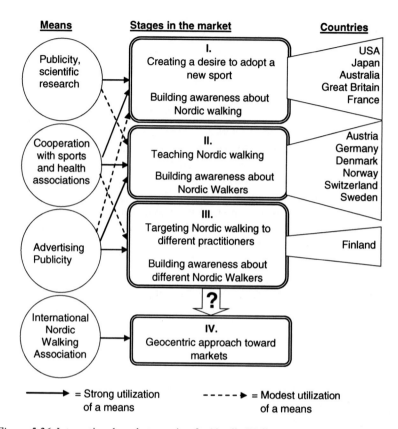

Figure 5.26 International market creation for Nordic Walkers.

can be very diverse, people coming from different organizations. In some places it's personal trainers, in some it's physiotherapists" (Senior vice president, Exel). These instructors then organize training through associations, community fitness centers and sports clubs. As in Finland, learning Nordic walking in groups also seems to be the trend abroad: it overcomes the shyness factor of being alone on the street using walking poles.

In general, this stepwise market creation has worked rather well in the Nordic and German-speaking countries. However, as competitors are increasingly entering the market with their imitations, this strategy of first promoting the sport and then the equipment could be too slow in new markets such as Japan and the USA, and a *more geocentric approach* may be needed in the future. The International Nordic Walking Association (INWA) was founded in 2000 to address this need and to promote Nordic walking around the world. It is becoming increasingly involved in coordinating the first-stage activities, i.e. creating the desire to adopt the new sport and building awareness.

Foreign markets accounted for more than 80 percent of Nordic Walkers' turnover in 2007. Future prospects seem bright, since the aging population in the industrialized countries is likely to foster increased demand. Despite the growing competition, the Exel Nordic Walker brand, together with the large instructor network and inventor image, enjoys a significant competitive edge in the markets in which it already has a firm foothold.

The role of cooperation in assuring the success of Nordic Walking

Both The Sports Institute of Finland and The Central Association for Recreational Sports and Outdoor Activities conveyed feedback from customers to Exel, which led to some post-launch modifications to the walking pole. For example, asphalt paws were added in 1999, and further improved straps and asphalt paws were introduced in 2003. A significant novelty product was launched in 2006 when Exel introduced the Calometer (an intelligent Nordic Walker pole that measures distance traveled, effort used and time spent, for instance). Exel considers continuous product improvement to be important for maintaining its commercial lead.

Ever since the launch, Exel has been getting direct feedback from its customers. The Internet has proved to be a particularly efficient form of communication. All feedback is taken into account in the product modification. "There have been a few strange ideas, but for the most part, where product development is concerned, the ideas have already been considered by our people and it's more a case of getting a kind of confirmation of them" (Senior vice president, Exel). Overall, Exel seems to have benefited considerably from *external linkages*. Cooperation has been mainly *diagonal*. At the idea-generation and development stages, the company cooperated mainly with The Sports Institute of Finland, which provided further market knowledge. At the launch stage, the input from The Central Association for Recreational Sports and Outdoor Activities was very important in promoting the innovation on the market. The joint interest in fostering sports activities motivated these organizations to cooperate with Exel. The cooperation seems to have worked rather well, and it still continues. Although the company regards both organizations as valuable cooperators, it has actively increased its internal capabilities in the development and marketing of the innovation over the years. Similarly, although both organizations understand Exel's efforts to increase its own capabilities, they wish to participate actively in the development of the pole and in the marketing of the sport in the future, since they have the direct contact network with masses of people practicing Nordic walking regularly. All these parties have been extremely *enthusiastic* about the sport and are committed to creating a large market for it. They all believe that it is not a fad and that it is here to stay: "What's been great is that we've realized that it didn't end up as a fad, but that it's become more like a lifestyle thing" (Senior vice president, Exel). "Well, it's really a product for everyone. It's a product aimed at the masses" (Secretary general, The Central Association). "I'd say that Nordic walking is here to stay. It doesn't become outdated. It's like walking, skiing, swimming or bicycling, or Nordic walking – that's it" (Head of physiological testing, The Sports Institute of Finland). The success of Nordic Walkers has been almost life saving for Exel: it has created enthusiasm and bright prospects for the future throughout the organization. The company was awarded first prize in the 2000 Innofinland contest, the aim of which is to promote significant Finnish innovations.

Furthermore, the Nordic Walker brand created a platform for Nordic Fitness Sports – a concept that was launched in 2003 and encompasses both summer and winter sports that activate upper-body muscles and provide the same kind of health benefits. In some cases, it promotes brand- and image-related issues (e.g. cross-country skiing became Nordic Fitness Skiing), whereas in others it involves the creation of new sports. For instance, Nordic blading is a high-speed refinement of Nordic walking that involves inline skating and specially designed, angled, Nordic Blader poles. The concept also incorporates Nordic snowshoeing, meaning walking with snowshoes and poles and thus getting exercise that is similar to Nordic Walking. This new variant is arousing increasing interest among various textile and footwear manufacturers who see it as an opportunity to develop and market sportswear to a big new target group, sedentary people.

5.5.7 *Customer-related proactiveness during the Nordic Walker development process*

In sum, although athletes had been using skiing poles on training walks, it could be said that the ideation for Nordic Walkers, special poles for fitness walking targeted on the masses, began in the mid-1990s when Exel started to consider ideas for a new sport. It took only five years to move from ideation to the mass market. The development process is illustrated in Figure 5.27.

Customer-related proactiveness seemed to vary along the developmental process (see Figure 5.28). The market opportunity was mostly *anticipated* at the idea-generation stage, when the firm was systematically trying to invent new

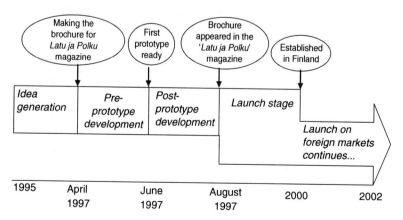

Figure 5.27 The innovation development stages of Nordic Walkers.

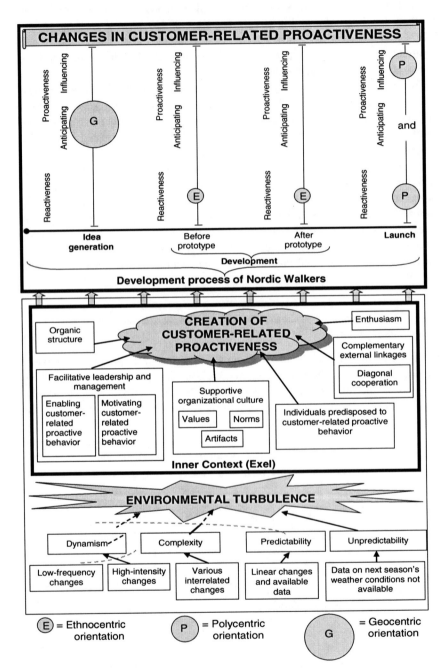

Figure 5.28 The change in customer-related proactiveness during the development of Nordic Walkers.

products. It analyzed the global trends and came to the conclusion that a new product related to walking might be successful. The feedback from the head of physiological testing in The Sports Institute of Finland gave further assurance that there might be a demand for walking poles. Since the analysis focused on global trends, it could be concluded that the orientation scope was *geocentric*.

However, at the development stage, the firm was able to use its long experience of ski-pole development. Ordinary ski poles were tested for walking and then extensively modified according to the feedback. Thus, it was possible to *react* to the customers' needs during the whole development process. The people who tested the ski poles and, later on, the walking-pole prototypes were Finnish, thus the orientation at this stage was *ethnocentric*.

During the launch, the firm was actively influencing its customers and created a market for both the new sport and a new device. Customers were made aware of, motivated and taught to practice Nordic Walking. Customer requirements were taken into account after the launch as well, and the pole has been further modified based on this feedback. It could therefore be concluded that both customer-related *proactiveness and reactiveness* were high at the launch stage. Since foreign markets have been approached individually, the orientation has been *polycentric*, which was necessary because the new sport required intensive awareness creation and teaching in each country.

When Nordic Walkers were created, the company environment was *somewhat turbulent*. Even though it was changing slowly, the changes were significant and interrelated, which in turn increased the turbulence. On the other hand, these changes were also well known and predictable, which then tended to decrease it. There was one unpredictable source of environmental turbulence that particularly affected Exel, a producer of winter-sports equipment, however, and that concerned the coming weather conditions.

The inner context in Exel was rather favorable to the creation of customer-related proactiveness. Its organic organizational structure and culture supported proactive behavior toward customers, which the leadership seems to have both motivated and enabled. In general, the people working in the organization could also be considered predisposed to customer-related proactive behavior. In this case, the complementary external linkages played a significant role both in the gathering of knowledge about customer needs and in changing those needs. The most important cooperative partners have been The Sports Institute of Finland and The Central Association for Recreational Sports and Outdoor Activities. In addition, various health and other sports associations have been particularly enthusiastic about the innovation and have contributed to the customer-related proactiveness. The role of enthusiasm is highlighted in that it is included in the framework.

In contradiction to the theory, building up customer-related proactive capabilities was not particularly emphasized in Exel. However, they may have been built up unintentionally as a by-product of the conscious recruitment of sporty employees.

Another difference to the theoretical framework is that liquid financial assets were not particularly stressed either. There seems to have been enough finance to develop and market the product in Finland, although the international launch might have been more extensive if there had been more finance. Still, it cannot be concluded that additional finance would have increased customer-related proactiveness per se, since it was always necessary to create the market by actively influencing customers. Additional finance would probably not have influenced the international scope of proactiveness either. Furthermore, a geo-centric orientation would not have been possible since there was a need to set up instructor networks in each country, and the network structure depended on country-specific circumstances.

5.6 Spyder

5.6.1 Spyder as a radical innovation

Equipment has traditionally played an important role in guaranteeing safe diving. Back in the 1950s, a professional diver from England contacted the Finnish compass manufacturer Suunto and told them that he had used their compass underwater. He suggested some modifications, on the basis of which Suunto developed their first underwater compass. However, divers need to be able to operate in three dimensions, and so in the 1970s, Suunto launched a range of diving instruments that could calculate depth as well as direction.

Advances in computerization in the 1980s enabled the development of the first diving computers. Suunto was among the first on the market and launched SME in 1987, a diving computer that was capable of calculating water pressure and temperature and providing information on how to conduct a safe dive. Traditionally, divers have leaned on manual calculations in estimating how long they could stay at a given depth. These calculations were complicated because they spend time at several depths, and each level requires a separate calculation. The diving computer replaced the need for manual calculations and provided more bottom time: divers did not need as wide a safety margin as when they were relying on their own calculations. This diving computer could be seen as a precursor of the wristop version. Suunto launched the world's first wristop diving computer based on sensor technology, the Spyder, in 1998 (see Figure 5.29).

Commercial success came very quickly, which helped the firm to confirm its position as the market leader in diving computers. It also led the way toward the creation of a new category: wristop computers, which steadily grew in import-ance to become the main product category. As a result, Suunto achieved annual growth rates of 35–40 percent (see Figure 5.30).

The *technological novelty* of Spyder is in the combination[29] of a diving com-puter and a chronometer in the form of a wristwatch. Suunto was able to combine existing technology in a new way and fit it all into a smaller case.

When the Spyder came onto the market, it provided new *substantial benefits* to the adopters. Although it is the size of a wristwatch, it provides divers with

Figure 5.29 The Spyder product.

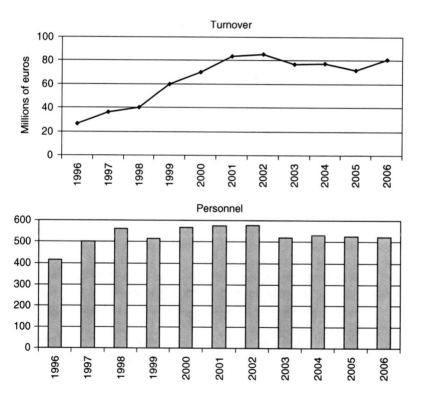

Figure 5.30 Turnover and personnel of Suunto Ltd 1996–2006 (Amer Sports Corporation Annual Reports 2000–2006; Rantasalo email November 2, 2004).

information on current and maximum depths, dive time, no-decompression time, and ascent rate, for instance. It has audible and visual alarms that warn divers when they are surfacing too quickly. It also has a memory function that tracks dive information – data that can later be downloaded to a personal computer and converted into graphic displays. It also has conventional watch functions, such as a calendar, a clock, a stopwatch and an alarm clock.

Consequently, Spyder *changed the behavior* of adopters. It permitted hands-free operations during the dive. It is light and easy to wear; still, it provides divers with the safety of conventional diving computers. They can thus concentrate more on diving, which makes the time under water more enjoyable.

However, it is not only the diving behavior that Spyder has changed. As the first diving watch that could be utilized all the time as an ordinary watch, it was the first of Suunto's *life-style products*. People who were practicing diving and were willing to express this kind of lifestyle were now able to do so by wearing a diving computer on their wrist (Kotro 2000: 4; Kotro and Pantzar 2002). The life-style product features were particularly emphasized in the later versions, especially in wristop outdoor computers.

In addition to the above, Spyder has certain features that seem to have influenced its development and marketing. First, it is targeted at wealthy people who are able to spend time and money on leisure: when it was introduced, it cost around US$1,000 and required a PC for the full benefits to be enjoyed. Second, its twofold characteristics (computer and watch) presented challenges in terms of design, marketing and retailing. Third, the product quality is of paramount importance, since the diver needs to be able to entrust his or her life to it: Suunto's slogan "replacing luck" suits this product rather well. Fourth, the market for diving equipment is rather small and dispersed around the world, which makes the business inherently global.

5.6.2 *Suunto Ltd – the developer of Spyder*

Suunto was established in 1936 when its founder, Tuomas Vohlonen, invented the liquid-filled compass. Since then, the focus of the firm has been on the development of measurement devices. Until the 1940s, the demand for compasses by the Finnish army was the mainstay of its sales, but when World War II ended and the Army's need for compasses decreased, Suunto had to target its products on the international market. Hence, the firm internationalized rapidly at a relatively early stage. Nowadays, it is one of the largest manufacturers of compasses and precision instruments in the world. It has been acquired several times over the years, and after a period of diversification, it is now again concentrating on sports instruments. It was taken over by the Amer Group, the sporting-goods supplier, in late 1999. Amer has given Suunto a broader financial and R&D base and access to larger distribution and marketing channels.

When the idea of Spyder was born, in the mid-1990s, Suunto belonged to Sponsor. Listed on the Helsinki Stock Exchange in 1995, it exported 97–98 percent of its production, enjoyed a 25 percent share of the world's diving-computer market and was the second or third biggest player on the market: the home market comprised only 0.3 percent of its diving-computer sales.

In 1995, the organizational structure of Suunto was *organic*. There were 12 people working on the development of a wristop diving computer. The small size of the team fostered open communication, and R&D was strongly connected to marketing. "The group was so small that almost all projects could be monitored around the coffee table, so openness was evident" (Technical director).

> In a way, product development was taking care of a lot of the functions that are usually in marketing's domain. The strength was that we got important end-user information without some middlemen filtering it. We were also able to interpret it correctly and make decisions fast, and we knew that if we wanted to add something, it would cost us this much money and take that much more time for development. It was definitely a strength to have a compact organization.
>
> (Director of Dive, Specialty and Military)

Consequently, the organizational structure facilitated the information exchange between marketing and R&D. Decisions were made quickly and without hierarchical boundaries. Control was rather informal and stemmed naturally from the normal collaboration in the everyday work:

> No one was able to work on his own for very long, so in a way the development and controlling the development was quite natural. Not through any formal, fancy processes, but through informal ones. We didn't have an organizational chart that would've said exactly how it goes. Somehow we were very much aware of the whole of what we were doing. [...] When we had problems then we did discuss things critically as well and people experienced it as a difficult situation. I wouldn't describe it as a very light and easy job.
>
> (Technical director)

The recruitment policy favored people who had diving as a hobby, which seems to indicate customer orientation,[30] although not necessarily customer-related proactiveness. Training was minimal due to the lack of time: "The pace of work was so hectic, that, although you should have time for training ... The attitude was that you'll learn best by doing it" (Director of Dive, Specialty and Military). Thus, building up customer-related proactive capabilities was not particularly emphasized in this case, nor was customer-related proactiveness, since the focus was rather on the development of project fluency in general. Although at the time there was no formal incentive-payment system, good work was rewarded through salary increases and career advancement. Moreover, the rapid growth of the firm also motivated the employees.

However, as the organizational structure fostered cooperation and knowledge sharing between marketing and R&D, it could be said that *customer-related proactiveness was enabled* in the firm. Consideration of customer needs was considered natural in such a small organization, and the culture emphasized the creation of benefits for diving enthusiasts. "Well, I'd say that bringing the benefit [to the customers] was what we were aiming at, it was definitely something that was at the core of what we were doing" (Technical director). The importance of understanding the needs and requirements of the customers was highlighted many times in the interviews and in the communications. It was also pointed out that when users do not know what their needs are, these needs must be demonstrated to them. The feedback from customers was also seen to have played an important role in the development work:

> The feedback that comes from the customers and end-users has always been extremely essential to us, whether it's positive or negative. Personally, I tried to emphasize the right kind of humility when communicating with customers and end-users. On the other hand, there's understanding the special needs of the sport, either through your own scuba diving hobby or the experiences of our test divers – it gave us the chance to evaluate the significance of the comments we'd received.
>
> (Director of Dive, Specialty and Military)

Consequently, it seems that proactive behavior toward the customers was *valued* throughout the organization, and customer focus could be considered a *norm* that was both *intense and crystallized*. It was also demonstrated in the *artifacts*, since the customers' desires were accentuated, as in the construction of Internet discussion forums.

The people working on the development and marketing of diving computers were rather young and unprejudiced. The director of Dive, Specialty and Military described them as open-minded and goal-oriented individuals, who were proud of what they were doing. Since many of them had diving as a hobby, they could anticipate the needs of other divers, and given the fact that they were developing something completely new, this knowledge gained through experience proved to be especially useful. Hence, even though they could not gather feedback from customers because they could not afford the information to leak out, and since the customers did not know "what they would like to wear on their wrist," they were still in rather good touch with customer needs. They were also keen to work in a firm in which they could develop a product they could use in their own leisure time. Thus, it could be concluded that these people were rather *predisposed to customer-related proactive behavior*.

5.6.3 Environmental turbulence

Traditionally, the market for field compasses has been very stable: as the director of Dive, Specialty and Military put it: "Nothing major ever happens there overnight, not even in a year or ten years!" The diving-compass market was changing in the 1980s, however, due to the launch of diving computers. In this situation, Suunto was actively searching for new market opportunities. It was

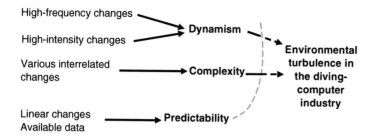

Figure 5.31 Factors contributing to the environmental turbulence in the diving-computer industry (the degree of unity of the arrows illustrates the degree of influence of the factors).

known that the popularity of diving was increasing, and in terms of diving equipment, it was the demand for diving computers that was growing most steeply. Hence, increase in demand was continuous and predictable.

The diving-computer market was becoming saturated in the middle of the 1990s, and the *competition* was getting very hard. The sales of Suunto's diving instruments stagnated in 1996: competitors were launching new products, and in order to maintain market share, the company had to decrease its prices. The biggest competitor was Uwatec, which had more financial resources to devote to aggressive marketing, whereas Suunto chose to concentrate more on R&D.

Computerization and *component technology* were advancing rapidly, which made it possible to pack more and more features into smaller and smaller products. Thus, the *dynamism*, at least in terms of technological change, seems to have been rather high: there were considerable and frequent advancements. Furthermore, the changes were many and interrelated, which increased the environmental *complexity*. All these technological changes seemed to have increased the turbulence of the diving-computer industry (see Figure 5.31).

However, the turbulence was decreased by the predictability of the changes: they were linear and data on the trends toward smaller and more multifunctional devices were readily available. Suunto was able to follow what was going on and combine its existing technologies in a novel way.

5.6.4 Idea generation

At the beginning of 1995, the CEO and some other key Suunto personnel were engaged in visioning novel ways to handle the toughening competition. The market leader, the Swiss firm Uwatec, launched a wrist computer incorporating wireless pressure-transmission capabilities, i.e. it could display tank pressure and calculate remaining air time based on current gas consumption. Importers of Suunto products urged the company to imitate this product:

> Of course, all of our distributors' first reaction was "make something similar, you've got to do it. You won't survive in this, you've got to make

it!" And we thought about that and analyzed it … Basically we then had two options, whether to make a copy and be sort of a "me-too manufacturer", and we knew that that technology included uncertainties, that it might not work in all kinds of circumstances, which later turned out to be true. On the other hand, why should we do what some competitor is doing? We realized we could do it our way and, despite the fact that the distributors were telling us what to do, we decided to create something of our own.

(Director of Dive, Specialty and Military)

In order to create "something of their own," the people in Suunto anticipated the future needs of the customers and tried to devise prospective new products. There were two interesting instrument groups. The first comprised diving computers that were attached to the wrist but were so large and specialized that they could only be used during the dive: the size of this market was rather limited. Second, there was the category of expensive diving watches, i.e. ordinary waterproof watches that were somewhat useful underwater and were also used in everyday life as indicators of a diver lifestyle: the market here was much larger. Consequently, the idea of creating a new product by combining features of these two product categories emerged:

That if we could combine the two products, the "diving only" side of a diving computer and on the other hand the characteristics of a watch and the status side of it all … It would be pretty easy to justify it to people. That someone wouldn't buy it just for the status value, but that he can invest in a diving computer and then use it otherwise, as well, and that way show to the world that he's a diver.

(Director of Dive, Specialty and Military)

This had two advantages. First, the new product was not likely to cannibalize Suunto's own traditional diving computers[31] but was likely to win customers from the top end of the traditional market for dive watches, and second, Suunto's traditional competitors were less likely to react, and prospective new competitors could be surprised. Thus, people in Suunto were confident in the new idea:

We had a feeling, a strong intuition, that by marrying up an expensive diving watch and a good, high-quality and high-performance diving computer, it would just have to work! From the information we'd gathered from around the world, we thought this was the direction things were going and that it'd give us more possibilities in other fields, later on.

(Director of Dive, Specialty and Military)

Communications with the Japanese importer convinced Suunto about this idea. A Japanese firm, Seiko-Epson, had already launched a watch-size product that could calculate depth in 1990. Suunto's aim was to launch a product that would include more advanced features and be more computer-like. The new computer seemed to provide an opportunity for combining Suunto's traditional mechanical know-how with electronics, thus strengthening its product development capabilities and opening avenues for the development of other watch-like computers. The watch-like computer was also seen as an important tool for building up the Suunto brand image, since it was a visible item that could be used to enhance sportiness:

> Whether it's golf or diving or whatever you're enthusiastic about, you only do it for a few hours a week. The point is that a product you only wear or hold in your hand while you're pursuing your hobby, other people rarely get to see it. One reason we developed this product was that typically people always have a few items with them wherever they go. With women, it's a handbag with their purse and their watch. With men, it's their wallet and their watch. So, if the watch is of a particular brand then everyone's going to see it all the time. And if the product works the way you expect it to, then that's really the best way to build a brand.
>
> (Director of Dive, Specialty and Military)

In sum, at the idea-generation stage, the opportunity in the market was *anticipated* based on Suunto's long experience in the field and on speculation about future needs. This was confirmed in discussions with the Japanese importer, which was keen on visioning future demand. Since Suunto had a long history of international operations, it was rather natural that it considered this opportunity to be global. Hence, the scope of anticipation was *geocentric* at this stage.

5.6.5 Development

The development of Spyder began at the end of 1995. Despite the fact that Suunto was able to utilize its previous know-how of precision-instrument manufacturing, the development work was still demanding because the precision-mechanics technology needed in watchmaking was new to them:

> It was a major step for us, as all the rest of them have been so much easier to make. As we don't really have the sort of right know-how in Finland when it comes to the fine mechanics of making a watch, or even the know-how to build the shell, we had to learn it all ourselves – it would've been a much easier job for a Swiss company. But luckily we had some good people who realized just how little they knew, which was fortunate, as then you don't make mistakes as you acknowledge that you have to get more information. We understood enough about the mechanics to understand how little we knew about these watch-sized gadgets.
>
> (Director of Dive, Specialty and Military)

The necessary skills and knowledge were obtained by establishing contacts with subcontractors in the watch industry. Some development work was outsourced, although Suunto's main aim was to increase its in-house know-how. Since the company believed that it could be developing a breakthrough product, the development work was kept very secret. *Before the prototype development, Suunto relied on the anticipation of market needs*, which was supported by communicating with a carefully selected small group of organizations:

> They were based on the discussions we'd had with those importers we trusted. [...] There were a few subsidiaries that knew. Not even all of the subsidiaries knew that we were working on something like this. The Japanese importer knew a little, and in the USA they had some clue about it.
>
> (Director of Dive, Specialty and Military)

> Well, the importers did express some wishes, but it was more about us going back to them and asking 'would this be OK?' and that 'we're working on this new model that has these kinds of things in it – what more would you want from it? Are these good things to have in it?' The ball was more in our court, because usually we made the product before anyone even asked for it, because that would've meant that we were already late.
>
> (Technical director)

Due to the structure of the organization, the developers had direct contacts with local importers, which facilitated the anticipation of customer needs:

> The product development had a lot of direct contacts. We didn't even actually separate marketing and product development, the people working in product development spoke to different importers directly about ideas, especially to those who'd been able to forecast in the past. We were working very closely with them, which meant that some of the hierarchy was dropped. But it requires a lot of the product-development people that they sort of have to be commercially minded and outgoing to get the clients to talk and describe things in the right way.
>
> (Director of Dive, Specialty and Military)

The first prototype was ready in October 1996. It was mechanically sound, but the software still needed considerable development, some of which was outsourced. Secrecy was maintained *after the prototype* development, although Suunto was able to collect feedback from some subsidiaries (especially in Germany) and importers. Furthermore, since Japan was considered an important market, the local importer was used for testing the new equipment. The feedback was useful:

> It was especially from Japan, and well, from other places too, that we got some [good ideas], because we were openly communicating to some of our trusted clients what we were doing and from them we got feedback about

what it could include and what it shouldn't. But this was a difficult product because we had to keep a very low profile about it as it was the first of its kind.

(Director of Dive, Specialty and Military)

Hence, at that time Suunto was able *react to the importers' perceptions of customer needs*. It was nevertheless important to try to pick out the signals that really reflected the end users' needs from the importers' feedback:

We're not here for each other, nor are we here for the importers. We're here for the end-users. Of course, in some cases the needs of the end-user and the importer and the retailer can coincide, but in the end it's really the needs of the end-user that make all this go round. That's what you've got to learn to understand.

(Director of Dive, Specialty and Military)

Thus, communication with retailing and subsidiary companies was demanding and also required diplomatic skills. It was not only a question of secrecy: another reason for this kind of feedback collection was that the product was so new that it was not possible to conduct market research. However, the need was also manifested in customer communication:

I can tell you a funny story that happened around the time we were about ready with this, when it [Spyder] was about to be launched in three or four months. This Swedish guy gave me a call and told me that he'd got some of our Diving-only computers and that they were really good and had we ever considered making a diving computer the size of a watch. He was adamant that it'd have a massive market. Of course I had a hard time squirming at the other end of the phone, muttering something about something might be launched in the near future ... I just couldn't tell him. I just said that it's terribly nice when people get in touch with us and that it's very important for us to get feedback from users and that these are things we're looking into at the moment. I'm sure that guy is now going around Sweden saying that he invented it and that Suunto manufactured it in three, four months. But it's just great to get that sort of feedback, it gives us more confidence that we're on the right track.

(Director of Dive, Specialty and Military)

The developers were particularly enthusiastic about the development of the new product. They were aware that they were making something that was unique in Finnish industry, as well as internationally. Enthusiasm was reflected in the willingness to work long days and in the commitment to solve the problems that were emerging during the development process. The post-prototype stage was lasting longer than expected, however, due to delays in the software development. This was a stressful period for the developers:

That final phase was really stressful and frustrating for everyone. It was hard to keep it under control when there were always new, say, software problems and so disappointments and all that. It got delayed again and again. It was a challenge to keep people motivated even after a lot of difficulties when the project managers and the people themselves were tired of it. But the team spirit was good, so...

(Director of Dive, Specialty and Military)

Product quality in a diving computer is a matter of life or death for the diver, and the developers themselves also tested the product by diving. Before the launch it was thoroughly tested in severe conditions.

Suunto invested almost FIM 10 million (€1.6 million) in the development of Spyder, and Tekes provided around 20 percent of the development costs. Surprisingly, the scarce liquid financial resources were later regarded as beneficial for the overall product development:

Well, in a way they [financial resources] were limiting [development] but on the other hand it might have been good at the time that it was necessary to look for simple solutions ... The scarcity led to sensible choices. If you think of it that way, then the scarcity didn't do us any harm, quite the contrary.

(Technical director)

The Technical Research Center of Finland (VTT) and various component and software subcontractors also contributed to the development, but *although this cooperation had an important role in the technical product development, it did not seem to increase customer-related proactiveness.*

The Spyder was designed in conjunction with Muotoilutoimisto Linja Oy. The product was small in size, with a display area as large as possible. Suunto's experience with diving enthusiasts played an important role in the design, which was modified by Linja and Suunto to suit the preferences of American consumers. As a result, the traditional Finnish simplicity was replaced with ornamentation and more use of the color of gold. The cooperation with Linja was successful, and it still continues.

Since Suunto was developing a product for the global market, it made every effort to take into account the different diving cultures around the world: "We tried to make an all-round product by combining things from different clients' perspectives, putting all these characteristics in the product and then you might've had things that were useless to Americans but maybe important to the Japanese" (Technical director). Thus, the computer display can be modified according to the user's own wishes. For example, the Japanese tend to make most use of the calendar and the time, whereas in the United States the most appreciated feature tends to be the "total number of dives" display. International aspects of the market were also taken into consideration in choosing a global name for the product, and Spyder was chosen on the basis of feedback from certain subsidiaries and key importers. Consequently, the orientation of both pre- and post-prototype development could be called *geocentric*.

5.6.6 Creating the market

It was important for Suunto to get the product to the market as quickly as possible in order to ease the cash flow and to benefit from first-mover advantage. The introduction was carefully prepared, and the launch material was ready well on time. Expectations of the soon-to-be-launched Spyder were high in spring 1997, although it was acknowledged that "almost anything could happen" with a high-tech product. Indeed, the problems with the software development delayed the introduction until autumn 1997, which meant that some of the launch-related marketing tasks were carried out too early:

> In terms of marketing, we acted too early. Firstly, what happens is that you lose money and secondly, something that was originally positive turns negative when people go to a store waving a wad of money in front of the manager asking to buy the product. Then the manager says he hasn't even seen the product. When that happens often enough with the same customer, or even several different customers, they go "Bugger off, I never want to sell your products again!" We came very close to that, it didn't quite go as we planned it. It got delayed too much, considering what marketing activities we had going on. But as it was the first one of its kind, then it's somehow understandable. We should've been more careful, though. We tried to be, but then the deliveries just got more delayed than they should've.
>
> (Director of Dive, Specialty and Military)

Thus, the launch stage began with setbacks. Since the importers in various countries were the key players in the launch, it was first necessary to arouse their interest.

Convincing the importers

At first, it was difficult to convince the importers about the benefits of Spyder, since there was no point of comparison. It was introduced to them at a big meeting that was held in a hotel in Helsinki:

> We just told them that "when you come over here to Finland, we're going to present you with something revolutionary." And the way we'd done it was that we had a magician with a proper magician's hat and all sorts of tricks. He was making things vanish and, finally, he pulled the Spyder out of his hat. It was fun looking at the expressions on these people's faces, they were clearly disappointed. They didn't really understand its value ... Well, some did, maybe 10% went "hey, this is really great!" About 90% didn't

understand what we were talking about. And one guy even sounded a bit desperate when he shouted, "Is it a watch or is it a diving computer?" He sounded like he wanted us to help him out.

(Director of Dive, Specialty and Military)

Suunto reassured the importers by describing the status features of the product and reminding them of the big volume of the watch market. It was further emphasized that the new product was a novel concept that did not threaten their existing range. Although the traditional sellers of expensive diving watches were retailers of timepieces, the Spyder wristop diving computer was first offered exclusively to diving-equipment retailers, as they were considered the natural channel: selling this kind of specialty product required both time and skill, as customers were rarely familiar with it or with all its functions.

It is typical of the diving-equipment business that retailers have a big responsibility for marketing. Suunto provided them with marketing material, which they either used as it was or modified to suit their circumstances better. Occasionally, financial marketing support was also given. The marketing budget was very limited however, and therefore they tried to make the product look so good that "it would sell itself". Heavier financial investments in marketing would not necessarily have guaranteed success any sooner:

It's hard to say whether it would've gone quicker, maybe. But it's difficult to say whether it would've taken more money than we had, in terms of results ... In my opinion, you can't always say that it's clearly the truth, that if you'd put more money into it, then it would've made more profit.

(Technical director)

The existing distribution network facilitated the Spyder launch, although the novelty of the product facilitated the upgrading of the retailing network:

Rather than having to knock on people's doors and ask whether they'd like to sell our products, it turned out the other way around and everyone was knocking on our door telling us they wanted to sell our products and we could just pick the best ones from the top.

(Director of Dive, Specialty and Military)

However, as the existing network was rather good and well established, there was no need for big changes. Certain features of the market for diving computers influenced the scope of the launch, and this is discussed in the following section.

International scope of the launch

Spyder was developed for a global launch. The main markets for diving computers are the United States, Germany, Great Britain, Italy, France, Japan, and

Australia. Suunto started the launch in Japan, where it presumed the demand to be considerable and where cooperation with the local importer was very close.

> If you take the most quality-conscious and demanding customers in the world then the first ones on the list are the Japanese, then a long, long gap and then the Swiss and the Germans. But the Japanese are in a league of their own. They can find every single problem a product has, but on the other hand, if something's been explained in the manual, that in a certain situation the product does this and this then it's not a problem to them. If you can make products good enough for the Japanese then you won't have problems anywhere else either.
>
> (Director of Dive, Specialty and Military)

Even though Suunto aimed to standardize its products and marketing efforts as much as possible, the marketing had to be adjusted to certain environments: in particular, the advertising, product colors and the name (Spyder became Spider) were changed to suit the Japanese market. After Japan, the United States was approached, being the biggest market for diving computers (40 percent of the world market), and both the volumes and the competition were particularly high.

Different diving cultures made entry into some countries more difficult than into others. It proved most difficult to sell the idea of a wristop diving computer in countries such as the United States and the Netherlands, where divers were used to having computers on consoles. It was thus necessary to focus on young divers who were free of prejudice. Another challenge in marketing the product to various countries was the difference in PC systems and software. Since the PC interface was an important feature, Suunto had to answer various questions related to its usability with various computer systems.

To conclude, the geographic scope at the launch stage seems to have been mainly *polycentric*: although Suunto was aiming for global standardization, it had to modify its approach to suit the different markets.

Reaching the customers

The creation of awareness was facilitated by the fact that Suunto was already known among divers and importers.

> We could do what Nokia is doing today, so we didn't need to push the product very much. We could just tell our most important clients [importers] when we were certain it was coming. There would then just be an immediate pull for the product so that we were able to sell the initial production in any case.
>
> (Director of Dive, Specialty and Military)

The diving-equipment manufacturing community is rather tightly knit, and regular participation in trade fairs is necessary in order to create awareness of new products. Spyder featured in various fairs, of which the DEMA (Diving Equipment and Marketing Association) Show organized annually in the United States was the biggest. Given its small size compared to diving-only computers, visibility was a challenge in both trade fairs and stores. This led to the use of enlargements in the displays and to the placing of advertisements in various divers' magazines. The advertisements highlighted safety, innovativeness and user-friendliness, and their role was primarily to create awareness and visibility.

Since diving can be dangerous, the use of *role models* has been an important selling feature. At one time, there was one global role model, Jacques Cousteau. After his death in 1997, there were only local opinion leaders who were diving instructors, professional divers and serious participants. Without the endorsement of these key people, the product would not have been credible in the eyes of the main market, i.e. occasional divers. Diving instructors in particular have been very influential in this respect:

> In all of these countries, the significance of the diving instructor is enormous. What he recommends, and what he likes … There aren't many sports in the world where the instructor's opinions count so much, parachute jumping maybe … Because when people go on the course, people instantly realize that if they do things differently from what the instructor says then their lives can be in danger. The instructors are so respected that their words count a lot.
>
> (Director of Dive, Specialty and Military)

Consequently, Suunto introduced the Spyder Diving Instructor Model, which had a Diving Instructor Special logo. It was both donated and sold on discount to instructors, who then would wear it and by their own behavior convince beginners about the product: beginners usually believed that the equipment the instructor used was of good quality and were often willing to buy the same product for themselves. Furthermore, *word-of-mouth communication* seemed to suit this product particularly well:

> If you think of any product that you don't have with you all the time, then you can only tell other people that it's a brilliant product. When people ask what it's like, ask you to show it to them, you can't because you've got it in a bag at home. But this kind of product that you can have with you all the time on a happy user … well, you're a free salesman for it when you start to show the product off and can show other people your latest [diving records].
>
> (Director of Dive, Specialty and Military)

Thus, through word-of-mouth communication, the product was also, in a way, selling itself.

Demonstrations were important in teaching people how to use the product. Suunto created the world's first interactive demo on diving computers, which gave prospective buyers simulated experiences of its use. The demo was also available on the Internet:

> So the fact that you can just simulate with it what the product does in different diving situations or on the surface, we've had that from a very early stage. That's the concept our competitors have started to copy as well.
>
> (Director of Dive, Specialty and Military)

Simulation was important since it gave the diver practical experience of the computer display in dry conditions; hence, he or she could practice reading it before diving. It also encouraged the buyer to enjoy the full benefit of the device by teaching him or her to utilize all its functions. In particular, it was acknowledged that customers often do not understand the full potential of the device when they purchase a wristop computer, and they tend to discover its various functions and capabilities later on as they use it.

Since Spyder cost almost twice as much as a normal diving computer, it was known beforehand that the buyers would be wealthy. Moreover, since it was known that older divers tend to prefer diving without a computer (and to rely more on their experience), it was assumed that the buyers would be relatively young: they turned out to be older than presumed.

> Seventy to eighty per cent of the people who bought the Spyder in the USA were really wealthy, over-45-year-old men. [...] People who are a bit older need reading glasses and then they prefer these bigger screens. But there was the flipside, that "OK, it's a small screen for a diving computer but it's a big screen for a watch."
>
> (Director of Dive, Specialty and Military)

At first, Spyder seemed to attract technology enthusiasts in particular: "Amazingly many were very technology-orientated people, people who are well-off, working for, say, IBM or Microsoft, these kind of electronics or software people. They really love products like these" (Director of Dive, Specialty and Military). Although most buyers were divers, some people bought the product because of the status features or because they wanted to convey a diver-like image. Furthermore, some non-divers bought it because they felt that they could fully rely on its waterproof characteristics.

Reacting to customer requirements

It was considered very important to accumulate the knowledge gained from customer feedback very actively:

> Speaking of communication, I think it's important that a company is interactive and especially always when a new product is involved. That you don't just passively wait and see how the product will do, but that you try to

find out immediately how it's doing. Because what happens is that in the beginning you're sorting the distribution channels and everything looks great, but then in, say, three, four months time the performance of the product really decided how much more you'll sell of it and you have to monitor the situation. [...] Communication is terribly important, and it has to be active.

(Director of Dive, Specialty and Military)

In practice, feedback from customers was collected both indirectly via the importers and directly via direct customer contact through e-mail, for example. It was easy to collect feedback from technology enthusiasts, since they were keen to comment on the product via e-mail. Furthermore, Internet discussion groups of diving enthusiasts provided an increasingly important feedback channel: "And you've got several for divers! When you follow them you can see directly what people think about your product, and competitors' products, for example improvement suggestions ... They're extremely valuable as information sources, especially for smaller companies" (Director of Dive, Specialty and Military). These discussion groups were significant word-of-mouth communication channels through which divers around the world could discuss their experiences. When word-of-mouth communication is conducted on the Internet instead of on the streets, the firm feels that it has at least some chance to turn negative experiences into positive ones:

It's been better not to interfere with it so that you stay invisible. In some cases if someone's had really bad experiences then you can approach the person directly and say, "If you've had some problems, then get in touch or send the product back to us."

(Director of Dive, Specialty and Military)

Some post-launch *modifications* were made to the product, the biggest of which was the lengthening of the battery life: although the product proved to be safe, the prototype-testing period was too short[32] to reveal the problems with the big power consumption of the batteries. These problems were manifested later at the launch stage, and they were fixed then. The color options were also changed. Originally, Spyder was available with various ring colors, but the choice was later reduced since Suunto noticed that the large variety of options slowed down the sales and was not even considered essential by the customers.

Based on the feedback gathered from Spyder's customers, a new model, Stinger, was developed. It included new software and new graphics and was launched in 2000. Mosquito and D3 followed later, and the company began to diversify from wristop diving computers into the outdoor wristop-computer market in general.

Establishing itself in the wristop computer market

The sales of Spyder began well and doubled in the second year compared to the first year. Suunto's traditional competitors were slow to react at first, and new prospective competitors were open to surprise. Currently, the competition in wristop diving computers is getting harder as competitors launch rival products: "The manufacturers of these expensive diving watches eventually realized that we'd penetrated their market and they had to react and manufacture similar products to the Spyder" (Director of Dive, Specialty and Military). However, Suunto has succeeded in increasing its market share and it is assumed that the demand for wristop computers will continue to grow. There is increasing interest particularly in countries in which console-attached computers used to be favored. Furthermore, as divers are traveling more and more to dive abroad, they tend to be increasingly interested in owning small diving computers, i.e. wristops. The company has become the biggest manufacturer of diving computers with a market share of 40 percent, and in 2006, the readers of the *UK Diver* magazine voted it the Diving Brand of the Year. Suunto has targeted the medium- and high-priced segments, although the market for diving equipment is limited. It is estimated that there are between three and five million active divers in the world.

Spyder was the first in Suunto's Outdoor Wristop Computer series (see Figure 5.32). Its electronic components formed the basis of the first wristop field computer, Vector, which was launched in 1998. Since then, special wristops have been developed for golf, marine, alpine-skiing and snowboarding enthusiasts. Regardless of the sport for which they are designed, all these computers provide the same level of information and the same kind of usability, and reflect the same basic philosophy of supporting performance. A wireless wristop computer incorporating mobile Internet access has been developed in conjunction with Microsoft and was launched in 2004. The outdoor wristop-computer market is very large and offers substantial potential for growth.

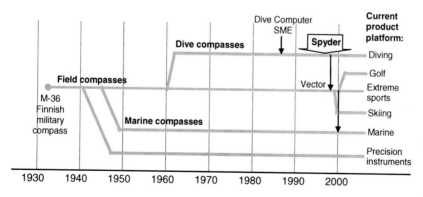

Figure 5.32 Spyder's impact on the evolution of Suunto's product lines (adapted from Lähteenmäki 2002).

There are currently around 70 people working in R&D in Suunto. As the organization has grown, it has become more challenging to keep employees motivated and enthusiastic. Around 98 percent of its turnover is generated abroad, its main markets being the United States, Europe and Japan. Prospective future markets include those in which consumers are wealthy enough to spend on their leisure time and to own PCs, which are often required in order to fully benefit from the equipment.

5.6.7 Customer-related proactiveness during the Spyder development process

To conclude, the innovation development of Spyder, from idea to the established product, took four years. The different phases of the process are illustrated in Figure 5.33.

Customer-related proactiveness appears to vary along the Spyder development process (see Figure 5.34). The market opportunity was *anticipated* at the idea-generation stage based on Suunto's long experience in the field and on speculation about future needs. The diving-equipment industry is inherently global, since the market is small and dispersed around the world, and the company naturally considered the market opportunity to be global. Hence, the scope of anticipation was *geocentric* at this stage. Furthermore, it was considered necessary *to add the role of the industry to the framework.*

The geographic scope was still *geocentric* at the development stage. The product was developed for the global market, and different diving cultures around the world were taken into account throughout the process. Before the prototype was ready, Suunto tried to *anticipate* market needs by communicating with a carefully selected small group of organizations. Believing that it was inventing a breakthrough product, it kept the development work very secret. Secrecy was maintained after the prototype development, although Suunto was able to collect feedback from its network of test divers, selected subsidiaries and

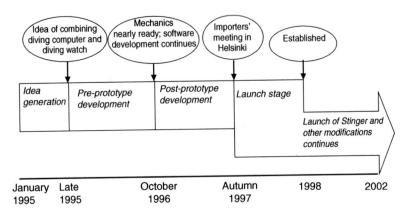

Figure 5.33 The innovation development stages of Spyder.

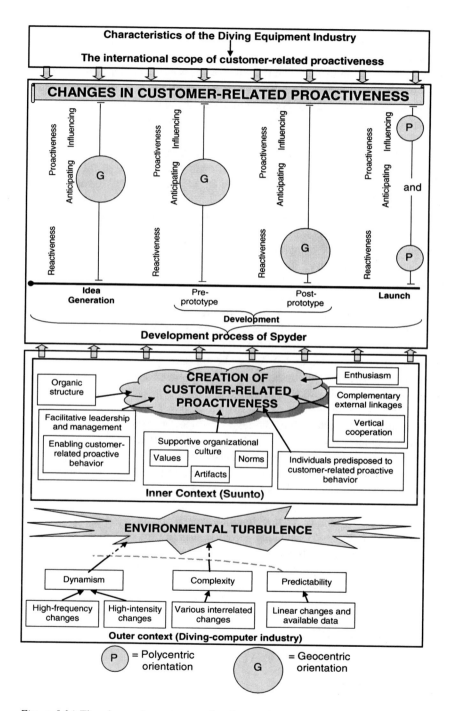

Figure 5.34 The change in customer-related proactiveness during the development of Spyder.

importers. It was thus able to *react to the importers' perceptions of customer needs*.

During the launch, Suunto first had to convince the importers of the value of the new product and then to help them to create awareness among customers. In order to reassure the customers, it was first necessary to get local diving instructors to accept it. Special demonstrations were arranged in order to give customers simulated experiences of using it, thus Suunto was actively *influencing* customers during the launch. The company also *reacted* to customer requirements post-launch and modified the product accordingly. Spyder was launched globally, but in order to respond to the different requirements of different customers, the marketing was adjusted to local circumstances. Thus, the approach at this stage could be called *polycentric*.

When Spyder was developed, technological changes in the environment (especially advancements in computerization and component technologies) created dynamism and complexity and thus also increased *turbulence* in the diving-computer industry. However, the predictability of these changes was likely to diminish the turbulence perceived by the firm.

Suunto's inner context seems to have fostered customer-related proactiveness. Its organizational structure was organic, and its culture emphasized this kind of behavior. The leadership seems to have promoted proactiveness. Furthermore, individuals working in the organization could be considered predisposed to customer-related proactiveness. People were also *enthusiastic* about the innovation, and since this enthusiasm seems to have increased proactiveness toward customers, it is added to framework.

Furthermore, it seems that during the whole development process, the close relationships with certain subsidiaries and importers promoted proactiveness toward customers. Thus, in this case, vertical cooperation has been particularly emphasized as a facilitator of proactive behavior.

It seems that the Suunto management was building up and motivating customer-orientation in general and hence proactive capabilities were not as strongly emphasized as suggested in the theoretical framework of the focal study. Another difference is that liquid financial assets were not considered particularly important in enhancing proactiveness toward customers, and it was even suggested that an abundance of these assets could have encouraged the launch of an unfinished product.

6 A modified framework of customer-related proactiveness during the development process

This chapter presents a cross-case analysis of the innovations studied. The building blocks of the initial framework are analyzed separately by comparing the cases with each other. Disparities and similarities are discussed, and the chapter ends with the presentation of a modified framework.

6.1 Degree of proactiveness

Customer-related proactiveness seemed to vary during the process of developing the radical innovation in all of the cases studied. The idea-generation stage was rather similar throughout (Figure 6.1), and *market opportunities were anticipated*. The idea generation for Nordic Walkers and ePost Letter involved a systematic search for new market opportunities. In the cases of Nordic Walkers, ePost Letter and DyNAzyme, the firms relied on their previous experience of customers, on the basis of which they anticipated that there might be a latent growing need for the kind of innovation in question. As far as Hi-Fog and Spyder were concerned, the stimuli came directly from the customers (Hi-Fog) or the retailers (Spyder), who urged the firms to "do something." In the case of Hi-Fog, the customer firm had a problem that it wanted solved, whereas with Spyder the retailer urged the developers to imitate the competitors, although after thinking about this they decided to develop something totally different. Thus, in these two latter cases, the need statements were also very general, and yet they formed the basis on which the firms had to deduce and anticipate what the market really wanted.

Consequently, influencing the market, which has been highlighted in the literature (e.g. Rogers 1983: 138–139; Veryzer 1998a), was not noticeable in these cases. It may be that past literature on radical innovations has tended to concentrate on industries that conduct basic research, whereas of these cases, only the firm that created DyNAzyme is engaged in such activities. However, this reasoning may not be adequate. First, even the idea generation of DyNAzyme relied on anticipation. Second, the idea for the innovation in the pilot study (the Domosedan case) was generated through basic research and even created opportunities; in other words, it influenced the market at the idea-generation stage. The firm was anticipating the latent needs of the market, and by trying to

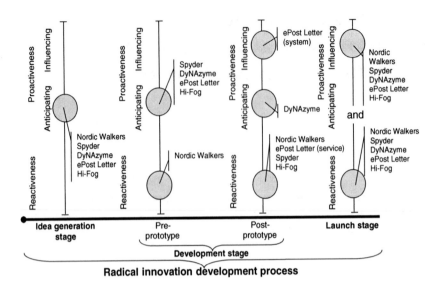

Figure 6.1 Customer-related proactiveness in the cases under study (Sandberg 2007a: 263).

respond to these needs it started to develop a blood-pressure drug for human beings. This development project did not turn out as expected, but surprisingly, *by chance*, it led to an idea for developing an animal sedative instead. Hence, the creation of a radical innovation that would actually open up opportunities on the market was not planned and was rather the result of serendipity[1] and the initiative of developers who were capable of seeing the new opportunity and capitalizing on it (Sandberg and Hansén 2004). This study thus shows that *opportunity creation at the idea-generation stage may not be as common as suggested in the literature.*

The cases indicated that it is not only the implementation of ideas (cf. Levitt 1963: 81) but also their creation that seem to require discipline and rather systematic anticipation of the emerging market needs. Hence, even at the very early stage of the processes described in the above cases, there was a presentiment of who the prospective customers would be, and there were attempts to anticipate what their needs might be.

During the development stage, most cases followed the pattern suggested in the literature (e.g. O'Connor 1998; Veryzer 1998b) in terms of reacting to the expressed needs of the customers after the prototype development. Hence, *customer-related proactiveness decreased during the development stage.*

The developers of Nordic Walkers, however, were able to react to the requirements before the prototype production because they were able to gather information from people practicing with skiing poles. Furthermore, since at that time there was no hint of the coming success of Nordic Walkers, it was possible to gather feedback rather openly, without worrying about the competitors. Consequently, *it seems that, since reactiveness tends to be more reliable and more cost-efficient than proactiveness, it is natural that firms prefer it, provided that it is feasible and can take place without jeopardizing their competitive advantage.* The adaptation of the innovation to customer needs early in the development phase increases its chances of success (McKenna 1989).

Following the prototype development, the developers of DyNAzyme relied on their field experience and continued to anticipate market needs. According to them, it would not have been possible to gather feedback from customers at that stage, since the suitability of polymerase products for gene technology can only be assessed after the ready enzyme has been launched: polymerases are used in the customers' own expensive biotechnical trials, and they have to be sure that the results are not affected by inferior quality.[2] Hence, *sometimes the characteristics of the market and/or the product may not permit anticipation even though the prototype is ready.*

An interesting way of behaving was evident in the ePost Letter case. After the prototype development, the developers started to directly influence other national postal operators and to create awareness of the new system. However, at the same time, they were modifying the service according to the customers' wishes. Hence, the end customers influenced the development of the innovation, whereas interest in it was aroused among the customers of the system. This could be explained by the need of developers to get the service launched quickly in order to bring in revenue, and there may not have been time to take the wishes of the national operators into account. Another reason may lie in the characteristics of the postal-service industry. National operators were used to developing their own systems and were not even considering buying a system from abroad until it was persistently offered to them.

There was both proactive and reactive behavior at the launch stage in all of the cases, which is in line with the theoretical framework. The creation of a market for a new innovation began by influencing the customers. The careful preparation of the customers (i.e. awareness building, education and trials) was particularly emphasized, and in each case the innovation was also modified after the launch according to the feedback received from customers.

6.2 The international scope of proactiveness

The international scope of customer-related proactiveness did not seem to vary as much as the degree of proactiveness (Table 6.1). In two of the cases, Hi-Fog and ePost Letter, the scope remained the same from the idea generation to the launch stage and *seemed to be rather predetermined by the field in which the firm operated.* The shipping industry, for example, is inherently global, and the geocentric scope of Hi-Fog development emerged naturally. On the other hand, postal services are traditionally in the hands of national operators, and therefore the scope of the ePost Letter service was ethnocentric throughout the development process. The scope of the ePost Letter system was polycentric, however, given the strong position of the national operators and the heterogeneity of their

systems. The influence of the industry structure was also apparent in the
DyNAzyme and Spyder cases, whereas it was not particularly accentuated in the
Nordic Walkers case. Roth and Morrison (1990) and Yip (1991), among others,
have emphasized the impact of the industry on the international orientation of a
firm. Thus, since the type of industry seems to have such a strong effect on the
international scope in all cases except one, it is added to the modified framework.

In the DyNAzyme case, the orientation was geocentric in all of the stages, but
polycentric influences were also evident at the launch stage, whereas with
Spyder, the orientation was geocentric until the launch stage when it became
polycentric. Hence, *it seems that, even though the needs of the market may be
anticipated or reacted to geocentrically at the development stage, influencing
them at the launch stage may require more targeted, polycentric orientation.*
This finding is in line with those of previous studies (e.g. Chryssochoidis and
Wong 1996; Hu and Griffith 1997: 119–122).

In the Nordic Walkers case, there was a shift from the geocentric (idea-
generation stage) through the ethnocentric (development stage) to the polycentric (launch stage). Hence, *although the firm anticipated worldwide
opportunities for the innovation at the idea-generation stage, the actual development of the innovation was ethnocentric.* This reflects the findings of previous
studies (e.g. Hedlund and Ridderstråle 1995; Lindqvist *et al.* 2000) that accentuate the local scope during the development stage.

At the launch stage, the firm actively shaped and influenced the needs of the
end market. These needs were influenced first in the home country and later in
the various foreign countries, again despite the fact that the market opportunities
were considered global. Customer-related proactiveness is expensive, and on the
global scale it demands even more resources (cf. Yelkur and Herbig 1996).
Therefore, it seems natural for the firm to consider very carefully how proactive
it should be. *The reality may push it to approach demand gradually, with an ethnocentric or polycentric orientation,* thereby accepting the danger of increasing
competition.

Table 6.1 The international scope of customer-related proactiveness in the five cases

	Idea-generation stage	*Development stage*		*Launch stage*
		Pre-prototype	*Post-prototype*	
DyNAzyme	Geocentric	Geocentric	Geocentric	Polycentric Geocentric
ePost service	Ethnocentric	Ethnocentric	Ethnocentric	Ethnocentric
Letter system	Polycentric	Polycentric	Polycentric	Polycentric
Hi-Fog	Geocentric	Geocentric	Geocentric	Geocentric
Nordic Walkers	Geocentric	Ethnocentric	Ethnocentric	Polycentric
Spyder	Geocentric	Geocentric	Geocentric	Polycentric

Furthermore, the study also demonstrated that, although *prototype develop-ment* played an important role in redirecting the degree of customer-related proactiveness during the development phase, it *did not have any effect on its international scope*. If the scope changed, it rather changed after the develop-ment stage, at the beginning of the launch stage. There are at least two reasons for this. First, there may be no need to change the scope during the development stage. It has been argued that experiential knowledge of customers is more bene-ficial during the development of new products for international markets (Li *et al.* 1999: 481), and it thus seems rather natural that the firm should rely on the inter-national scope with which it has gained such knowledge. Second, the develop-ment stage of a radical innovation is such an intensive and challenging effort in itself (cf. Veryzer 1998b: 146–149) that there may not even be the time or the resources to consider the adaptation of an international scope in the middle of it.

6.3 Fostering proactiveness

The analysis of the organizational factors influencing proactive behavior was rather difficult due to the retrospective nature of the study. Therefore, the find-ings presented in this chapter should be considered tentative, put forward with a view to inspiring further research rather than providing a thorough description of the organizational context.

In all the cases in question, *the organizational structure seems to have been organic* (Table 6.2). This kind of structure emerged rather naturally in small firms such as Marioff and Finnzymes, but even in Finland Post, which as a state-owned giant had a generally mechanistic structure, the team that developed the radical innovation was able to establish an organic suborganization within it.

According to the theoretical framework, leadership and management may facilitate the emergence of customer-related proactiveness through the building up of customer-related proactive capabilities and enabling and motivating this kind of behavior in the organization. Correspondingly, *the management was enabling and/or motivating the customer-related proactive behavior in the firms* under study. It was interesting to note that *building up customer-related pro-active capabilities was not accentuated in any of the cases*. This suggests that these capabilities were not particularly emphasized in the recruitment or training practices. However, upon closer examination it became clear that, even though they were not wittingly built up, they *might still have been created unintention-ally*. For example, in the Nordic Walkers' case, the firm was consciously recruit-ing sporty people who, according to the literature (e.g. Courneya and Hellsten 1998: 630; Reiss *et al.* 2001: 1143), tend to have similar characteristics to pro-active people. This kind of reasoning may apply to other cases, as well: after all, the characteristics of proactive people (e.g. conscientiousness and extraversion) often tend to be favored in recruitment (Robertson and Smith 2001: 449–450), even though proactiveness is not explicitly sought.

The organizational culture was supportive of proactive behavior toward the customers in all of the cases except DyNAzyme. In this case, it was strongly

Table 6.2 Organizational factors in the five cases

		DyNAzyme	ePost Letter	Hi-Fog	Nordic Walkers	Spyder
Organic structure		yes	yes	yes	yes	yes
Facilitative leadership and management	Building up customer-related proactive capabilities					
	Enabling customer-related proactive behavior	yes	yes		yes	yes
	Motivating customer-related proactive behavior					
Supportive organizational culture	Values		yes	yes	yes	yes
	Norms		yes	yes	yes	yes
	Artifacts		yes	yes	yes	yes
Complementary external linkages	Horizontal	yes	yes	yes		yes
	Vertical	yes	yes			
	Diagonal				yes	
Liquid financial assets		yes	yes	yes	yes	yes
Individuals predisposed to customer-related proactive behavior		yes	yes	yes	yes	yes
Enthusiasm		yes	yes	yes	yes	yes

oriented toward product development and scientific research, which is natural in the biotechnology field in which new emerging technologies may rapidly replace old ones (cf. Meldrum and Millman 1991: 48). Hence, even though the organizational culture did not support customer-related proactiveness to the same extent as in the other cases, it may be that the culture was still more supportive than in biotechnology firms in general. After all, the developer of DyNAzyme was also active in the retailing business, which inherently brought knowledge about customers and their needs to the firm.

As also suggested in the literature (e.g. Slater and Narver 2000: 12), all the firms in question utilized *complementary external linkages* to enhance their customer-related proactiveness. The danger of creating a potential competitor tended to limit the cooperation to the most reliable collaborators.

In contrast to the theoretical framework, however, *liquid financial assets were not considered particularly important* in enhancing proactiveness toward customers. It thus seems, as Nayak and Ketteringham (1986: 12) suggested, that in order to develop successful radical innovations, "the key is not big resources, but the right resources at the right time." This kind of reasoning is also supported in the study carried out by Gemünden *et al.* (2005: 371).

The role of *individuals predisposed to customer-related proactive behavior seemed to be important* in all of the cases, which again corresponds to the theoretical framework: there were certain key people who were both proactive and customer-oriented and who were able to stimulate customer-related proactiveness in their organizations. In the Hi-Fog and DyNAzyme cases, these individuals were the founders of the firm, which is in line with previous literature suggesting that founders of technology-based firms tend to have a significant impact on organizational behavior (Bruno 1989: 34; cf. Holt *et al.* 1984: 33).

However, these people were also key figures in conveying *enthusiasm* about the innovation. This enthusiasm spread through the organization, and it seemed to motivate people to create potential markets. Thus, it was an important promoter of customer-related proactiveness. This finding is similar to the results that Nayak and Ketteringham (1986: 19, 345) obtained when they analyzed various radical innovation development processes: they found that the spirit and emotion within individuals fuelled the creation of radical innovations.

In sum, the results of the case studies suggest the need to modify the a priori framework by removing two factors: the management's building up of proactive capabilities and liquid financial assets. On the other hand, it seems appropriate to add enthusiasm to the model, since its role was emphasized in all of the cases.

6.4 Environmental turbulence

The results of previous studies on the role of environmental turbulence in the formation of proactive behavior are contradictory (see Section 2.2.2). This study did not clarify the issue either, since the turbulence varied between the cases. Hi-Fog was developed in a turbulent environment, whereas in other development processes, although there were some big changes, their predictability decreased the turbulence.

Overall, the relevance of "environmental turbulence" as a concept could be questioned. It has been criticized by Mintzberg (1993: 33–37), for example, who suggests that people tend to be inclined to see their own age as more turbulent than past ones. Furthermore, he argues that the total environment of a firm is seldom turbulent, and that it rather has both turbulent and stable elements. He claims that, in fact, turbulence is only a change that strategic planning cannot handle (see also Mintzberg 1994: 204–209). It thus seems that *there is no need to include the environmental-turbulence dimension in the modified framework.*

It may also be the case that the operationalization of the concept "environmental turbulence" utilized in this study could not capture the phenomenon. Although it did not work here, however, it cannot be completely ruled out for future investigations based on these results. On the contrary, its utilization and operationalization in further studies would appear to deserve careful consideration.

6.5 Synthesis

Figure 6.2 presents the theoretical framework modified according to the results of the case studies. As suggested in the original framework, the degree of customer-related proactiveness changed during the process of developing the radical innovation. Although in most cases the changes were consistent with the framework, there was more variation between the cases, especially at the post-prototype development stage.

The international scope of customer-related proactiveness varied considerably between the cases, but within each one a change in scope was not as common as a change in degree. Since the industry in question seemed to influence the international scope, it was added to the framework.

The case studies also generated some modifications to the organizational variables: management's building up of proactive capabilities and liquid financial assets were removed, and enthusiasm was added. Furthermore, since the case studies gave as contradictory information as the past literature on the role of environmental turbulence, the outer context was removed from the framework.

The framework served its purpose rather well. It was useful in understanding customer-related proactiveness during one radical innovation development process, and it also enabled comparison between the cases. The utilization of the reactiveness–proactiveness continuums to describe the degree of proactiveness was also rather successful, as was the use of the ethnocentric, polycentric, regiocentric and geocentric (EPRG) typology to illustrate the geographic scope. The organizational variables were described on a relatively general level, since the retrospective approach limited deeper description. In a real-time longitudinal study, the framework could be more detailed in this respect and could allow the analysis of changes in the variables during the innovation development process.

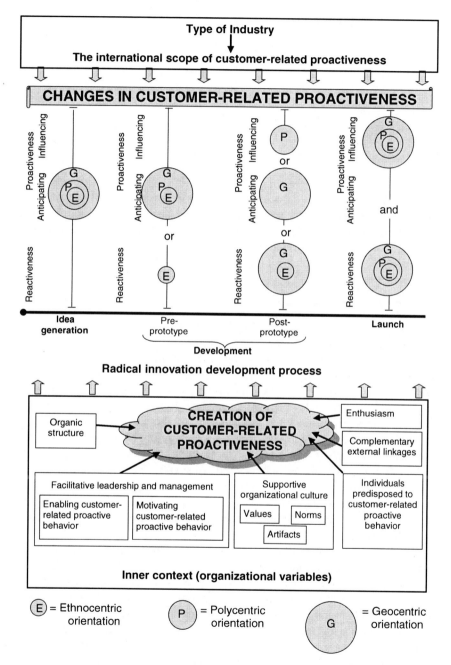

Figure 6.2 The change in customer-related proactiveness during the development of radical innovations (modified framework).

7 Conclusions

This chapter begins with an evaluation of the trustworthiness of the study. It then continues with discussion of the theoretical conclusions, arranged in the form of propositions and in terms of its contribution to prior theoretical knowledge. The managerial contribution is then assessed, and this chapter ends with a discussion of the methodological issues and suggestions for future research.

7.1 Trustworthiness

Various researchers have suggested different guidelines and criteria for evaluating qualitative research. Some (e.g. Silverman 2001: 225–255; Yin 1989: 40–45) have applied the same basic terminology that is utilized in evaluating quantitative studies, whereas others (e.g. Lincoln and Guba 1985: 290–327; Patton 1990: 460–493) have developed new terminology that, according to them, is more suited to the qualitative approach. However, although researchers may use different terminology, the contents of the various criteria seem to resemble each other. One of these criteria was therefore chosen as the basis for evaluating the trustworthiness of the focal study. Lincoln and Guba's (1985: 290–327) criterion has been widely used, apparently facilitating the extensive analysis of trustworthiness, and it was applied in this study. They suggested that, in order to evaluate the trustworthiness of a qualitative study, the researcher ought to consider the credibility, transferability, dependability and confirmability of the research.

Credibility refers to how well the constructions at which the researcher has arrived correspond to the multiple constructions of reality that the subjects of the inquiry have (Lincoln and Guba 1985: 295–296). In this case, the researcher aimed to increase the credibility of her study by familiarizing herself with the theoretical background and forming a theoretical framework before entering the field (cf. Miles and Huberman 1994: 17). Key concepts, such as customer-related proactiveness and radical innovations, were operationalized through the preliminary conceptual framework. The interview themes and questions were based on the theoretical framework and modified in the light of the expert interviews and the pilot study. Hence, every attempt was made to accumulate sufficient preunderstanding of the phenomenon.

There is always a risk that interviewees will attach different meanings to key concepts (cf. Ottesen and Grønhaug 2002). Here the researcher returned the case descriptions to the interviewees for verification to ensure that her understanding of the course of events corresponded to how the interviewees experienced the innovation–development process (cf. Lincoln and Guba 1985: 314; Patton 1990: 468–469).

Since the researcher had not been involved in or observed the actual innovation–development processes and had access only via interviews and secondary data, the credibility of the results may be threatened. However, there were abundant secondary data, and the selected interviewees had played key roles in the development processes, and this reduced the risk.

The difficulties inherent in interviews in terms of past time were also evident in the focal study. Generally, the interviewees remembered the innovation–development process and the actions taken during it surprisingly well, but when the interview moved to the organizational context, the responses clearly became more hesitant.[1] The researcher resolved the problem to some extent by asking some specific questions and by comparing the responses of various interviewees (cf. Collins and Bloom 1991: 26). Nevertheless, the limitations of the retrospective approach have to be acknowledged, and the analysis of organizational factors was consequently kept on a rather general level.

Leonard-Barton (1990: 250) strikes a note of caution in suggesting that interviewees may have particular difficulties recalling events they did not recognize as important when they occurred. This would apply to recollections of the idea-generation stage in this study, although it seems that the interviewees realized at a very early stage that they were developing something special in generating ideas for radical innovations, and recalled certain past moments very clearly. This is in line with the results of a study by Huber (1985) indicating that participants in organizational processes do not forget key events as quickly as one might imagine.

There is always the danger in interviews that the interviewees are not willing to reveal the real course of events (Collins and Bloom 1991: 28–29; Mason *et al.* 1997: 314). Golden (1992: 855) suggests that this may apply in particular to pet projects with which they are publicly associated and emotionally involved. Each of these radical innovation development processes clearly represented this kind of high point in the interviewee's career. The innovations had been awarded various prizes, such as "Innovation of the year" and "Innofinland." The interviewees referred to them as "my favourite kid," and many also referred to themselves as "the father of the innovation." Hence, there was clearly a danger of painting too rosy a picture of the course of events. Nevertheless, the respondents rather openly talked about the difficulties they had experienced during the process. This openness may, in fact, be a result of the time lag. Furthermore, the use of multiple respondents and data triangulation, i.e. the combination of the interview data with secondary data, also seem to have decreased the bias caused by nostalgia (cf. Huber and Power 1985: 175–176; Leonard-Barton 1990: 258).

The retrospective approach also has its advantages. The time lag may, in fact, clarify the process and increase focus on the essentials: it makes the informants remember and concentrate on the events and decisions they consider significant (Leonard-Barton 1990: 262).

Transferability refers to the extent to which the results of the study are applicable to other empirical or theoretical contexts (Lincoln and Guba 1985: 296–298). Lincoln and Guba (1985: 316) emphasize the fact that the researcher cannot specify transferability in qualitative studies, but he or she can, and should, describe the data in such detail that the potential appliers are able to judge for themselves. The rather detailed case descriptions that are included in this study provide the reader with enough information to enable them to assess the transferability to other contexts.

Tying the research to existing studies through literature reviews may also improve transferability in case studies (cf. Eisenhardt 1989: 544–545; Lukka and Kasanen 1995: 77–78). In this case, the theoretical framework used for the analysis of the empirical data, i.e. literal replication logic, seemed to increase the potential for analytic generalization (cf. Yin 1989: 38–40).

Dependability indicates how dependent the findings of the study are on the entity in question, and how much the inquiry itself affects the results (Lincoln and Guba 1985: 298–299). The influence of the interviewer on the results of qualitative studies is considerable: according to Patton (1990: 279), "The quality of the information obtained during an interview is largely dependent on the interviewer." This influence on the findings begins in the very early stages when the interviewer devises the questions. When they are based on the theoretical background, as in this study, the data obtained is theory-laden (cf. Olkkonen 2002: 111). A formal atmosphere and possible interruptions during the interviews may also distort the findings (cf. Eskola and Suoranta 1998: 90–95). In this case, there were certain interruptions (e.g. telephone calls or other employees coming into the office), but they did not seem to have a major impact on the information obtained.

Confirmability refers to how easily other researchers find it to confirm the research results (Lincoln and Guba 1985: 299–301). It is acknowledged that a researcher's selective perception inherently distorts the results, and hence some other researchers may extract other relevant issues from the data (cf. McKinnon 1988: 37–38). Here the researcher tried to enhance confirmability by writing detailed case descriptions and illustrating them with citations.

Furthermore, she has reported the research process in detail, so that readers are able to evaluate how the research was conducted and how the conclusions were drawn. The innovations in question are publicly known, and the names of the interviewees are openly reported. The main secondary sources are carefully documented in Appendix 3, so that the reader is able to trace them. This is also likely to increase confirmability.

This section described the trustworthiness of the focal study in terms of the criteria of credibility, transferability, dependability and confirmability. Both weaknesses and strengths were contemplated; most of the weaknesses apparently somewhat related to the retrospective nature of the research. Every effort was made to decrease these weaknesses during the research process.

7.2 Theoretical conclusions

7.2.1 Propositions based on this study

The study aimed to provide a framework for analyzing customer-related proactiveness. The framework is explorative, since although innovation development has been widely studied, and there has recently been a growing stream of studies on radical innovations in particular, this study seems to be among the first to link customer-related proactiveness with radical innovation development.

Given its explorative nature, the most significant theoretical implications are *the propositions arising from the modified framework*. The study revealed that the degree of customer-related proactiveness changes during the development process, and consequently, it is recommended that *future research on innovations should consider this variation rather than considering customer orientation in general*. This follows the lines of a study conducted by Larson *et al.* (1986), according to which the degree of proactiveness is related to various task and environmental factors (see also Luo 1999).

Proactiveness at the development stage varied between the cases but was similar during the idea-generation and launch stages. Hence, the following propositions are put forward:

- *Proposition 1: At the idea-generation stage, firms developing radical innovations anticipate the market opportunities.*
- *Proposition 2: At the launch stage, firms developing radical innovations prepare the market before the launch and react to the market requirements after the launch.*

Although the importance of international markets is generally recognized in innovation management, there seems to be a lack of research on how these markets are taken into account in practice. This study contributes to filling this knowledge gap by being among the first to link customer-related proactiveness, an international scope, and radical innovation development. The international scope varied between the cases, but was rather consistent within each one. Closer analysis indicated that the scope was influenced by the type of industry in which the firm operated.

- *Proposition 3: The type of industry influences the international scope of customer-related proactiveness.*

Thus, it seems that the stage of the innovation development process mostly influences the degree of proactiveness, whereas the type of industry seems to affect the geographic scope (Figure 7.1).

Furthermore, a tentative framework was put forward describing the organizational variables that seem to influence the creation of customer-related

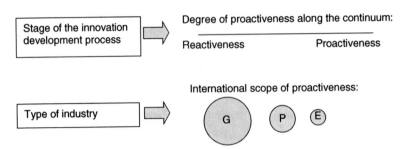

Figure 7.1 Influences on the degree and scope of customer-related proactiveness.

proactive behavior in a firm. The following five propositions are based on the theoretical background and the case studies:

- *Proposition 4: An organic organizational structure fosters customer-related proactiveness.*
- *Proposition 5: Managers foster customer-related proactiveness by enabling customer-related proactive capabilities and by motivating customer-related proactive behavior.*
- *Proposition 6: The organizational culture influences customer-related proactiveness.*
- *Proposition 7: Individuals involved in the process of radical innovation development, and who are predisposed to customer-related proactive behavior, foster customer-related proactiveness during the process.*
- *Proposition 8: Complementary external linkages foster customer-related proactiveness.*
- *Proposition 9: Enthusiasm among those developing the innovation fosters customer-related proactiveness.*

The role of environmental turbulence in customer-related proactiveness was also investigated. This varied between the cases and in fact did not seem to be useful in an explanatory sense. This supports the criticism put forward by Mintzberg (1993: 33–37). Since previous studies have also provided contradictory results regarding the role of environmental turbulence, *its utilization and operationalization in further studies ought to be carefully considered.* A recent study by Buganza *et al.* (2004), which recognizes the complexity and presents a matrix for analyzing turbulence dimensions via market and technology determinants, could provide a fruitful avenue for the further development of the concept.

Furthermore, the study contributed to the conceptual discussion on proactiveness, market proactiveness and customer-related proactiveness. Customer-related proactiveness was considered in the light of "anticipating" and "influencing," both of which were further operationalized in the interview themes. Hence, even though it may be difficult to empirically test the whole modified framework later by means of quantitative research, the mere operationalization of the proactiveness concept in this study is likely to serve further studies. This is of particular significance, since the latest research seems to emphasize the need for further studies on market-oriented proactiveness (e.g. Narver *et al.* 2004).

7.2.2 Contribution to prior knowledge

The main contribution of the focal study stems from the combination of customer-related proactiveness, the process of radical innovation development, the international scope of proactiveness and organizational factors. It is among the first studies to combine these streams of research.

The study elaborated the concept of proactiveness and introduced a way of describing a firm's proactiveness as a pattern. The utilization of this kind of pattern could also be of use to researchers *studying the strategic behavior of firms* as it could give a more detailed picture than a typology or a continuum (cf. Miles and Snow 1978; Vesalainen 1995).

Furthermore, the concept of customer-related proactiveness was clarified. This contributes to *the study of market orientation*, which has previously emphasized reactiveness (e.g. Kohli and Jaworski 1990), but in which more attention is being focused on the proactive side (e.g. Narver *et al.* 2004). Hence, this research is in line with recent requests by Kumar *et al.* (2000) and Slater (2001a) for researchers to pay more attention to proactive behavior.

The few existing studies on market proactiveness (e.g. Kumar *et al.* 2000; Narver *et al.* 2004; Slater and Mohr 2006) have treated it as a static concept. Firms have been considered either proactive or reactive, based on how they behave in relation to a multitude of divergent issues (see Figure 7.2). This study

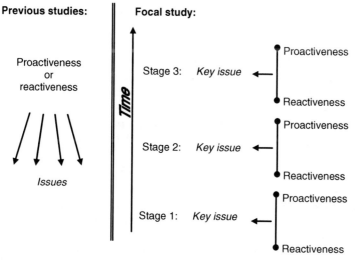

Figure 7.2 Proactiveness and reactiveness in previous studies versus in the focal study (Sandberg 2007a: 264).

introduces a process-oriented approach. The processual study of proactiveness is complicated by the time perspective, which is inherent in the very concept. Therefore, in this study, the proactiveness during different stages of the process was tied to certain key issues with regard to which the firm was either proactive or reactive. This enabled degrees of proactiveness to be illustrated through positions along the reactiveness–proactiveness continuums and a time-dimension to be incorporated.

The utilization of the key issues and reactiveness–proactiveness continuums could also be useful in other *longitudinal studies*, which are needed since proactiveness seems to be related to various task and environmental factors (cf. Larson *et al.* 1986: 397). Consequently, the process approach adopted in this research may encourage further longitudinal studies on the phenomenon.

This study also contributed to *previous research on the development of radical innovations* by illustrating how the degree of proactiveness toward customers changes during the process. Contrary to many earlier studies (e.g. Veryzer 1998a), this research demonstrated that anticipation seems to play an important role at the idea-generation stage.

In terms of *international business research*, the focal study demonstrated that the ethnocentric, polycentric, regiocentric and geocentric (EPRG) typology is also suitable for illustrating the international scope of a firm's orientation and that this may well be combined with research on market orientation. Although Perlmutter, one of the developers of the EPRG typology, suggested this as long ago as in the 1970s (Wind *et al.* 1973), it has not been widely applied in this sense before.

Research in the field of international business has recently emphasized the role of rapidly internationalizing firms, or "born-globals." These firms are also often technology-intensive and, in fact, their internationalization may be based on radical innovation (cf. Hurmelinna *et al.* 2002). Thus, this research could also open up new avenues for understanding born-global firms.

Furthermore, the study indicated that the type of industry seems to influence the international scope of proactiveness. The globalization of industries may in the future foster a geocentric scope of innovation development in an increasing number of firms (cf. Hurmelinna *et al.* 2002). Since proactiveness in this context demands resources, it is likely to create particular challenges for small and medium-sized enterprises. Hence, in the future, this may increase the importance of complementary external linkages such as cooperation and networks in fostering global proactiveness.

As far as *organizational factors are concerned*, this study brought up the importance of enthusiasm in increasing customer-oriented proactiveness. Enthusiasm has not featured in previous studies describing factors influencing proactiveness, nor has it been particularly emphasized in previous research on innovation development. Consequently, its role in future studies of both proactiveness and innovations could be more carefully taken into account.

7.3 Managerial conclusions

The managerial implications of the study stem from the evident need for innovative firms to develop commercially successful products. These firms seem to be facing increasing pressure to enhance proactiveness toward customers. In fact, proactiveness requires substantial amounts of time and resources given the newness of the innovation. Geocentric proactiveness is even more demanding. *Firms ought therefore to consider carefully how proactive they should be, i.e. when it is appropriate to invest in customer-related proactiveness, and when it is not.*

It seems from the cases that, particularly at the idea-generation stage, firms anticipate the opportunities in the market. Thus, finding out what kind of products or services customers might need seems to be important at the very early stages of the innovation development process. The results of this study indicate that a systematic search for new market opportunities, and the firm's previous experience of customers, may generate not only incremental (cf. Trott 2002: 233–234) but also radical innovations.

Proactiveness is not always needed, however, as the cases indicate: during the development stage in particular, some of the firms were able to behave reactively toward the customers. The study also indicated that, even though a firm is developing a radical innovation for the global market, the geographic scope at the development stage may be more limited. It is worth noting, however, that when the innovative firm behaves reactively toward its customers and limits its international scope, it may put its first-mover advantages at risk.

The study also showed that firms tend to behave both proactively and reactively during the launch stage. They may prepare their customers for the coming innovation by raising awareness, educating potential customers and giving them opportunities to test the innovation. Furthermore, listening to customer feedback and modifying the innovation accordingly seems to have been widely used among the firms in question. However, all this demands resources, particularly in situations in which international markets are also approached from a polycentric perspective.

Innovative firms aiming at increasing proactive behavior toward customers need more knowledge *on how proactiveness can be fostered in the organization.* The framework created in the study may help them to focus on key issues. For example, it may inspire managers to consider how they could enhance enthusiasm within their own organization, and if resources are scarce, it may encourage them to use complementary external linkages.

Certain implications for policy making were also highlighted. It was shown that, even at the idea-generation stage, firms tend to anticipate the market opportunities, but because of their limited resources, this anticipation mainly rests on secondary sources and information gained from the media. Therefore, governmental organizations could support these activities by *providing abundant and good-quality information on future trends.*

Another issue worth suggesting is that organizations funding innovation development projects perhaps ought to pay more attention to *evaluating the organizational variables* of the applicant firms, since these seem to be important in fostering proactiveness. Funding decisions nowadays are rather tightly concentrated on innovation, whereas the developing and commercializing organization is left somewhat in the dark.

7.4 Methodological considerations and suggestions for further research

The initial framework for this study was first created on the basis of the theory and then modified in the light of multiple retrospective case studies. Its creation and modification increased understanding of customer-related proactiveness during the development of radical innovations. However, the study had certain limitations that create interesting avenues for further investigation.

First, given the exploratory nature of the study, further research aimed at developing a more thorough understanding of the complex and multifaceted issue of customer-related proactiveness would certainly be of interest. The propositions presented in Section 7.2.1 should be evaluated according to additional data. It would be useful to test the framework through a large quantitative survey. Quantitative research in this context is demanding, however, since the interviews in the case firms revealed that different people define the stages of the innovation development process in different ways. Thus, it would be extremely hard to ensure trustworthiness. Moreover, there are often many key people involved, which would make the targeting of the survey very difficult. The limited number of radical innovations also limits the possibilities for quantitative research. Hence, a more realistic avenue for further research may be in conducting more case studies.

Second, the cases studied comprised different kinds of radical innovations targeted on different kinds of markets. The only common ground was that they were all commercially successful radical Finnish innovations. Hence, in order to better understand customer-related proactiveness in innovation development, it would be extremely interesting to test the framework on different types of innovations: a study of *incremental innovation* would reveal to what extent developmental proactiveness is different from that in radical innovation.

Third, it has to be recognized that Finnish firms are used to looking for international opportunities rather actively, due to the small size of the home market. It would therefore be of *interest to study firms in larger countries* and to see how the size of the home country influences the international scope of customer-related proactiveness.

Fourth, it would be extremely interesting to test the framework with *unsuccessful innovation development* projects in order to see whether some kind of "pattern of successful behavior" could be detected.

Finally, the focal study could produce only tentative results regarding the role of organizational variables in the creation of customer-related proactiveness. In order to have studied these variables more thoroughly, it would have been necessary to conduct *a longitudinal real-time study and to collect data at various organizational levels*. However, the characteristics of radical innovations did not permit this kind of approach within one small research project. More extensive research projects would increase our knowledge of the organizational variables. One of the variables that has apparently been neglected in past studies on innovation development is *the role of enthusiasm*. It would be very interesting to study innovation-related enthusiasm in more detail in terms of what it encompasses, and how it could be conveyed within the firm between employees and between functions.[2]

8 Summary

The motivation for the study arose from the idea that firms developing radical innovations may be facing "a mission impossible." On the one hand, in order to succeed, they have to develop innovations that fulfill the needs of customers, while on the other hand, in the case of radical innovations, they may not even know who the prospective customers might be. Furthermore, even if the customers are known, they often do not recognize the need for a completely new kind of product or service. Hence, reacting to customer needs and wishes is often not possible. Proactive behavior toward customers, in other words anticipating and influencing them, may thus be the key issue in building the bridge between the innovation and the market.

The purpose of the study was to analyze the role of customer-related proactiveness in the process of developing radical innovations. This was pursued through the following sub-objectives:

1 to analyze what customer-related proactiveness in developing radical innovations encompasses;
2 to analyze the degree of customer-related proactiveness during the process of radical innovation development;
3 to analyze the manifestation of the international context in customer-related proactiveness during the development process;
4 to describe how customer-related proactiveness is created in firms that produce radical innovations.

The study began with a literature review, based on which a preliminary framework was created. The empirical part of the study comprised expert interviews and a pilot case study, followed by five Finnish case studies of radical innovation development. The framework was further modified based on the results of the empirical study.

The first sub-objective was *to analyze what customer-related proactiveness in developing radical innovations encompasses*. On the basis of previous literature, this was defined as either action based on information gathered about customers before their behavior has had a direct effect on the firm or action that deliberately influences and creates changes in customer behavior.

The second sub-objective *was to analyze the degree of customer-related proactiveness during the process of radical innovation development.* Past studies indicate that the degree of proactiveness toward customers' changes during the innovation development process, and indeed, the stage of the process seemed to influence the degree of proactiveness in the cases studied. The firms anticipated the market opportunities at the idea-generation stage and employed varying degrees of proactiveness at the development stage, but at the launch stage, they prepared the market beforehand and also reacted to market requirements afterwards.

The third sub-objective was to analyze the manifestation of the international context in customer-related proactiveness during the innovation development process. The international scope was determined in light of the EPRG model, which classifies orientation as ethno-, poly-, regio- and geocentric. It was found that the geographic scope varied between the cases but was rather consistent within each one. The type of industry seemed to have an important effect on the geographic scope of the firm.

The fourth sub-objective was *to describe how customer-related proactiveness is created in firms that produce radical innovations.* The results indicated that an organic organizational structure, enabling customer-related proactive capabilities and motivating customer-related proactive behavior by management, a supportive organizational culture, complementary external linkages, enthusiasm, and the presence of individuals predisposed to customer-related proactive behavior, all seemed to foster customer-related proactiveness.

The study was inspired by the question: "To what extent is it possible to act proactively toward customers when developing radical innovations?" It appears from the results that, although the market for a radical innovation develops along with its creation, it is not necessarily hidden from the firms developing it. On the contrary, these firms seem to be able to anticipate the opportunities in the market and develop radical innovations that fulfill the unarticulated needs of customers.

On the whole, the results of the study suggest that customer-related proactiveness is a multifaceted issue and should be taken into account in future studies. Calling a firm proactive, or saying that behavior during the innovation development process is proactive toward customers, may be nice and neat, but at the same time it may foster neglect of significant variation in behavior.

Appendix 1

Glossary of key concepts as defined in this study

Concept	Definition in the focal study
Anticipating	Detecting signals of change, following these signals and forecasting their coming changes and acting accordingly (Ashley and Morrison 1997: 48).
Artifact	Material or non-material thing that communicates information about the organization's values and behavior.
Customer	An end user or someone who influences or dictates the choices of end users. Thus, the concept of "customer" is extended to the end user's support group, management and distribution channel, for instance. (Kohli and Jaworski 1990: 4; Wah 1999: 19). Furthermore, the concept of "customer" includes both current and prospective customers.
Customer needs	Divergences between the customer's existing and the wanted situation (Kärkkäinen 2002: 6; cf. Holt *et al.* 1984: 8).
Customer-related proactiveness	Either acting based on the information gathered about the customers before their behavior has had a direct impact on the firm or deliberately influencing and creating changes in customer behavior.
Development	See "Development process"
Development process	A process extending from the idea generation to the development stage and finally to the launch stage.
Development stage	A process of putting a new idea in a physical and psychological entity that is expected to meet the needs of an audience of potential adopters (Rogers 1983: 139–140; Urban and Hauser 1993: 39).
Environmental turbulence	Dynamism, complexity, and turbulence of the environment (Volberda 1998: 191–195).
Ethnocentric customer-related proactiveness	Influencing and anticipating the coming market circumstances based on information gathered only from the customers of the home country (cf. Wind *et al.* 1973: 15).
Geocentric customer-related proactiveness	Influencing and anticipating the behavior of customers on a global scale.

Concept	Definition in the focal study
Idea	A concept, a thought, or a collection of thoughts (Trott 2002: 12).
Idea-generation stage	The stage that begins with searching for or picking up ideas for a new product and ends with the decision to go further with the product development.
Influencing	Creating changes in the environment.
Innovation	A commercially successful invention.
Invention	An idea that is converted into a tangible artifact (Trott 2002: 12).
Launch stage	The stage that starts with planning the launch and ends when the innovation has become established in the market.
Market orientation	The organization-wide generation, dissemination, and responsiveness to market intelligence, which includes both reactive and proactive behavior (cf. Kohli and Jaworski 1990: 3).
Market proactiveness	Either acting based on the information gathered about the market before the circumstances in it have had a direct impact on the firm or deliberately influencing and creating changes in the market
Market reactiveness	Responding to certain circumstances in the market after they have had a direct impact on the firm.
Norms	Shared expectations regarding what are appropriate or inappropriate attitudes and behaviors.
Organic organizational structure	An open, loosely controlled and informal organization that has a high degree of participation in decision-making and a low degree of hierarchical authority and where multidisciplinary teams are favored.
Polycentric customer-related proactiveness	Influencing and anticipating customer behavior in various countries individually (cf. Wind *et al.* 1973: 15).
Proactiveness	Either reacting to the symptoms of the coming change in the firm's environment or creating changes within it (cf. Johannessen *et al.* 1999: 118).
Radical innovation	A new product or service that requires considerable change in the customers' behavior and is perceived as offering substantially enhanced benefits and which is also technologically new.
Reactiveness	Responding to the environmental changes after they have had an impact on the firm (cf. Stacey 1993: 123).
Values	What is considered desirable (Shrivastava 1985: 105) and important (Ott 1989: 39)

Appendix 2
List of interviews, discussions and correspondence

Interviews

The Domosedan case (pilot case)

Kuussaari, Jukka, former Marketing Director of the International Division, Farmos, Salo 12.3.2002.

Lähdesmäki, Kai, former Vice President of the International Division, Farmos, Turku 13.3.2002.

Vähä-Vahe, Tapani, former Research Manager, Farmos, Turku 8.3.2002.

Expert interviews

Kankaala, Kari, Director, Technology Transfer, The Finnish National Fund for Research and Development (Sitra), Turku 10.5.2002.

Sipilä, Kari, Executive Director, The Foundation for Finnish Inventions, Helsinki 28.3.2002.

Case interviews

Akonniemi, Matti, Director, Letter & Direct Marketing Services, Finland Post Corporation, Helsinki 2.6.2003.

Forsberg, Arto, Export Manager, Finnzymes Ltd, Espoo 13.6.2003.

Heikkilä, Matti, Head of physiological testing, Coach, Sports Institute of Finland, Vierumäki 11.12.2002.

Hämeenoja, Marita, Customer Relationship Director, Finland Post Corporation, Helsinki 22.5.2003.

Jantunen, Tuomo, Secretary General, The Central Association for Recreational Sports and Outdoor Activities (Suomen Latu ry), Helsinki 2.12.2002.

Karihtala, Aki, Senior Vice President, Sport Marketing, Exel Sports Ltd., Vantaa 7.11.2002.

Koivunen, Jarmo, Former Director of the PEM unit; Former Managing Director of International Data Post (IDP), [retired], Helsinki 5.6.2003.

Lähteenmäki, Ari, Director: Dive, Specialty and Military, [employed by Polar in late 2003], Suunto Ltd., Vantaa 19.6. 2003.

Mattila, Pekka, CEO, Finnzymes Ltd., Espoo 8.4.2003.

Nikkola, Ari, Technical Director, Suunto Ltd., Vantaa 14.10.2004.
Siljander, Mirja, Product Manager, Communications Solutions, Atkos Ltd. (Atkos is part of the Finland Post Group), Helsinki 25.3.2003.
Sundholm, Göran, CEO, Marioff Corporation, Vantaa 24.10.2002.
Tuomisaari, Maarit, Manager (R&D), Marioff Corporation, Vantaa 24.10.2002.
Viljasaari, Jari, Development Manager, Finland Post Corporation, 5.6.2003, Helsinki.

Discussions and correspondence

Akonniemi, Matti, Director, Letter & Direct Marketing Services, Finland Post Corporation, 30.11.2004 (e-mail).
Alho, Jukka, President and CEO of Finland Post Corporation, 2.3.2003 (e-mail).
Helamaa, Erkki, Architect, Professor Emeritus, 10.11.2004 (telephone).
Himanen, Hanna, Marketing Assistant, Exel Sports Ltd., 2.5.2003 and 17.11.2003 (e-mails).
Holm, Maikki, Marketing Coordinator, Finnzymes Ltd., 8.4.2003 (face-to-face discussion); 29.6.2004 and 1.7.2004 (e-mails).
Huoso, Sari, Corporate Communications, Exel Oyj, 30.9.2003 (e-mail).
Hämeenoja, Marita, Customer Relationship Director, Finland Post Corporation, 2.9.2004 (e-mail).
Jantunen, Tuomo, Secretary General, The Central Association for Recreational Sports and Outdoor Activities, Suomen Latu ry, 30.10.2003 and 16.2.2004 (phone), 28.11.2003 and 3.12.2003 (e-mails).
Karihtala, Aki, Senior Vice President, Sports Marketing, Exel Sports Ltd., 12.11.2002, 24.4.2003, 7.11.2003, 13.2.2004 and 20.6.2007 (e-mails).
Koivunen, Jarmo, Former Director of the PEM unit; Former Managing Director of International Data Post (IDP), [retired], 5.6.2003, 5.12.2003, 12.12.2003, 2.3.2004, 5.3.2004 and 2.12.2004 (e-mails), 3.3.2004 (phone).
Manninen, Taisto, Testing Manager, Exel Sports Ltd., 2.12.2003 (e-mail).
Nummila, Hanna, Executive Assistant, Marioff Corporation, 24.10.2002 (face-to-face discussion).
Ollila, Sanna, Marketing assistant, Metorex Security Products Oy, 15.11.2004 (e-mail).
Rantasalo, Tuija, Accountant, Suunto Oy, 2.11.2004 (e-mail).
Siljander, Mirja, Product Manager, Communications Solutions, Atkos Ltd. (Atkos is part of the Finland Post Group), 10.11.2004 and 12.11.2004 (e-mails).
Skyttä, Vuokko, Managing Director, Atkos Ltd. (Atkos is part of the Finland Post Group), 7.3.2003 (e-mail).
Sorsa, Pirjo, Vice President (Finance), Marioff Corporation, 28.2.2003 (telephone), 1.4.2003 (e-mail & telephone), 4.4.2003 and 14.6.2007 (e-mails).
Stockmann, Sten, Product Manager: Diving, Suunto Ltd., 1.7.2004 (e-mails).
Sundholm, Göran, CEO, Marioff Corporation, 28.2.2003 (telephone).
Sunila, Timo, Product Designer, Fiskars Consumer Oy Ab, 7.12.2004 (e-mail).
Tulkki, Martti, Manager of Technical Services, Finnforest Kerto, 8.12.2004 (e-mail).
Tuomisaari, Maarit, Manager (R&D), Marioff Corporation, 4.4.2003 and 12.5.2003 (e-mails).
Viljasaari, Jari, Development Manager, Finland Post Corporation, 6.9.2004 (e-mail).

Appendix 3
The most important publications utilized to complement the case descriptions

The DyNAzyme case

Secondary documents

Beck, S. (1998) "High Fidelity PCR: Enhancing the Accuracy of DNA Amplification," *The Scientist*, 12, 1: 16–18.

Erill Sagalés, I. (2002) *High-speed Polymerase Chain Reaction in CMOS-Compatible Chips*, Barcelona: Universitat Autonòma de Barcelona.

Gene Technology Act, Ministry of the Environment www.miljo.fi/default.asp?contentid=30969&lan=EN (accessed 27.4.2004).

Hemmann, K. (2001) "Innovaation tuotteistaminen," *Biologisten funktioiden tutkimusohjelma Life2000 News*, 2: 5–6.

Inkinen, H. (1990) "Vuoden teekkariyrittäjien lähettämässä paketissa on restriktioendonukleaaseja," *DiA-kunta* 9: 32.

Mattila, P. and Pitkänen, K. (1989) "Finnzymes Oy – ideasta yritykseksi," *Kemia – Kemi* 13, 3: 265–267.

Miettinen, R., Lehenkari, J., Hasu, M. and Hyvönen, J. (1999) *Osaaminen ja uuden luominen innovaatioverkoissa*, Helsinki: Sitra 226, Taloustieto Oy.

Ojanperä, K. (1988) "Finnzymes Oy porautuu geeninsiirron entsyymeihin," *Insinööriuutiset*, 1988, 12: 4.

Ojanperä, K. (1989) "Finnzymes laajentaa koko molekyylibiologiaan," *Tekniikka & Talous*, 11.8.1989: 12.

Pavlov, A. R., Pavlova, N. V., Kozyavkin, S. A. and Slesarev, A. I. (2004) "Recent developments in the optimization of thermostable DNA polymerases for efficient applications," *Trends in Biotechnology*, 22, 5: 253–260.

Rabinow, P. (1996) *Making PCR: A Story of Biotechnology*, Chicago: The University of Chicago Press.

Radiaattori 12.5.2004, 10.03–10.43. Radio Programme. Interviewees Pekka Mattila, Peter Bredbacka and Mikko Koskinen. Editor S. Loikkanen. YLE Radio 1.

Swanson, D. (1999) "PCR Optimization: Building the Perfect Beast," *The Scientist*, 13, 4: 26–27.

Veikkola, V. (1992) "Finnzymes tekee Suomen kalleimpia tuotteita," *Kauppalehti Optio*, 17: 51–52.

Voitto+ (2007) CD ROM containing financial statements data of Finnish companies. Suomen Asiakastieto Oy: Helsinki.

Öster, H. (1997) "Finnzymes jättäytyi suosiolla patenttiriitojen ulkopuolelle," *Tekniikka & Talous*, 9.10.1997: 15.

Company documents

Finnzymes Internet pages: www.finnzymes.fi (accessed 7.4.2003 and 30.5.2007).
Mattila, P. (2002) "Finnzymes Oy – 15 vuotta suomalaista entsyymiteknologiaa," *The Zymes*, Customer Magazine of Finnzymes, November 2002, 9: 1.
Product Catalog 2001–2002. Finnzymes Ltd.

The ePost Letter case

Secondary documents

Elle, P.J. (1993) "Prisvinner i hard konkurranse om elektronisk post," *Post, Postverkets bedriftsavis*, 23: 3.
Fuller-Love, N. and Cooper, J. (2000) "Deliberate versus emergent strategies: A case study of information technology in the Post Office," *International Journal of Information management*, 20, 3: 209–223.
International Data Post Bringing Post Offices to the Internet Age. Case Study. Sun Microsystems. www.sun.com/service/sunps/success/case_studies/idp_cs2.html (accessed 8.3.2004).
J. Granfelt and J. Viljaranta, *Kilpailun lisäämisen taloudelliset vaikutukset postitoiminnassa* (1996) Helsinki: The Helsinki Research Institute for Business Administration.
Mackintosh, E., Mackintosh, I. and Willmott, R. (1999) *Hybrid Mail in the Third Millenium*, Kingsley: Mackintel Ltd.
Nikali, H. (1998) *The Substitution of Letter Mail in Targeted Communication*, Helsinki: Helsinki School of Economics and Business Administration.
Nikali, H. (2001) "Demand Models for Letter Mail in the Competitive Targeted Communication Market?," Draft, *Institut D'Economie Industrielle Second Conference on Competition and Universal Service in the Postal Sector*, Toulouse, France, 6–7.12.2001.
Norwegian IT-group acquires International Data Post (2001) Press Release 30.10.2001. PostInsight, Press Releases. <www.postinsight.pb.com (accessed 11.11.2003).
Pietiäinen, J.P. (1998) *Post for All: The Story of the Finnish Postal Service 1638–1998*, Helsinki: Edita.
"Pohjoismaat siirtyvät elektroniseen kirjeliikenteeseen" (1991) *Tekniikka & Talous*, 2.9.1991: 20
"Samnordiska IDP, marknadsledare inom ePost" (1993) *NPT Nordisk Posttidskrift*, 3: 10–11.
Uotinen, P. (1991) "Posti vie sähkökirjejärjestelmäänsä," *Kauppalehti*, 29.8.1991: 7.
Vennamo, P. (1999) *Pekka posti ja sonera [kertomus tosielämästä]*, Helsinki: Otava.
Viherä, M.-L. and Luokkola, T. (1993) *Tavoitammeko toisemme? Posti- ja teletoiminnan perusteet*, Helsinki: Opetushallitus.

Company documents

Asap, Customer Magazine of Atkos.
Atkos Internet pages: www.atkos.com (accessed 20.3.2003).
Finland Post Annual Reports 1990–2006.
Finland Post Internet pages: www.posti.fi (accessed 4.7.2002 and 16.6.2007).

Finland Post Renamed Itella on 1 June. www.itella.com/group/english/current/2007/
    20070531finlandpostrenameditellaon1june.html, 1.6.2007.
Postia, Customer Magazine of Finland Post.
Postiset, Employee Magazine of Finland Post.
Posts and telecommunications of Finland, 1980–1988, Official Statistics Finland,
    Helsinki.
Tietolinja, Employee Magazine of Posti-Tele.
Vakkala, M. (1992) *Introduction to ePost*. Video Presentation at the 8th World Confer-
    ence & Annual General Meeting of Technology Centers, June 10–12, 1992. Oulu,
    Finland.

The Hi-Fog case

Secondary documents

Arvidson, M. and Hertzberg, T. (2001) *Släcksystem med vattendimma – en kunskapssam-
    manställning*. Borås: SP Sveriges Provnings- och Forskningsinstitut.
Conventions. International Maritime Organisation. www.imo.org/Conventions/index.
    asp?topic_id=148 (accessed 28.1.2003).
Eriksson, G. (2001) "Många är på väg in i vattendimman," *Aktuell Säkerhet*, 6: 36–38.
Grant, G. (2000) "New European Water Mist Standard," *IWMA (International Water
    Mist Association) Newsletter*, 1: 4.
Korteila, M. (2001) "Patenttinikkarille työ on kiva harrastus," *Tekniikka & Talous*,
    8.2.2001: 10–11.
Lehtinen, L. (2002) "Pieni pisara on suurta palontorjuntaa," *Tekniikka & Talous*,
    23.5.2002: 4.
Maritime Safety. Finnish Maritime Administration. www.fma.fi/e/functions/safety
    (accessed 30.1.2003).
The Montreal Protocol on Substances that Deplete the Ozone Layer (2000) Nairobi:
    Ozone Secretariat, United Nations Environment Programme.
Orkoneva, O. (2006) "Keksintö kasvoi mereltä maalle," *Europa, Euroopan komission
    Suomen edustuston lehti*, 4: 26–27.
Pepi, J.S. (1996) "Water Mist Fire Protection Systems Come of Age," *BrandPosten*,
    15.12.1996: 10–11.
Puustinen, T. (2001) "Hyvin sammutettu," *Talouselämä*, 2.12.2001: 24–26.
Stewen, P. (2001) "Marioff Water Mist for Fire and Smoke Control," *Navigator*, 5:
    20–22.
Sundell, B. (2002) "Eldens betvingare," *Huvudstadsbladet*, 4.11.2002: 1, 8.
Valtonen, A. (2002) "Göran Sundholm – Tuhannen patentin mies, joka nauttii kaupan-
    teosta," *Tekniikan Akateemiset*, 5: 21–22.
Wallén, M.(2001) "Lite Brandsläckare lönsam som Nokia," *Dagens Industri* 13.11.2001: 14.
Wighus, R. (2002) "Water Mist Fire Protection," *Industrial Fire Journal*, September
    2002: 45–48.

Company documents

First Choice, Information bulletins of Marioff Corporation.
Imagine … HI-FOG (2001) Corporate video. Marioff Corporation.

Marioff Corporation Internet pages: www.hi-fog.com (accessed 10.1.2003 and 15.6.2007).
Press releases of Marioff Corporation.

The Nordic Walkers case

Secondary documents

Aaltonen, P. (2002) "Sauvakävely johtaa uusien lajien rynnistystä liikunnassa," *Helsingin Sanomat* 4.9.2002: C9.

The Central Association for Recreational Sports and Outdoor Activities, Internet pages: www.suomenlatu.fi. (accessed 14.9.2005 and 14.5.2007).

Church, T.S., Earnest, C.P. and Morss, G.M. (2002) "Field Testing of Physiological Responses Associated with Nordic Walking," *Research Quarterly for Exercise and Sport*, 73, 3: 296–300.

"Exelissä sauvoista vauhtia mastoon nousuun" (1999) *Tekniikka & Talous*, 14.10.1999: 13.

Haworth, D. (2002) "Helsinki: Nordic Walking anyone?" *Europe*, 421: 34–35.

Ingves, B. (2001) "Exel stavar snabbt framåt," *Finland levererar*, Finsk-svenska handelskammaren 3: 20.

International Nordic Walking Association, Internet pages: inwa.nordicwalking.com. (accessed 14.5.2007).

Klasmann, J. (2001) "Fit mit Stöcken," *Gesundheit, Das Magazin für Lebensqualität*, 4: 82–83.

Miralles, P. (2002) "Caminar con bastones; un deporte en auge en los paises nórdicos," May 2002. www.andarines.com/equipo/nort.htm (accessed 29.10.2003).

"Nordic Walking. Ganzkörpertraining mit zwei Stöcken" (2001) *Sport@all, Das Branchen-Magazin*, 3: 26–27.

Nummela. S (2003) *Sauvakävely 2003*, Espoo: TNS-Gallup Oy.

Nurminen, T. (2003) "Sauvojia satumaisissa maisemissa," *Kauppalehti Optio*, 17: 86–89.

Raevuori, A. (2001) "Sauvakävely – kansanliike," *Latu ja Polku*, 4: 30–31.

Siivonen, R. (2002) "Sauvakansan synty," *Ylioppilaslehti*, 89, 15: 10–11.

Thurwachter, M. (2001) "Beat inactivity with a (Nordic) stick," *The Palm Beach Post Journal*, 11.1.2001: 3E.

Trötschkes, R. (1999) "Die Stockgänger," *FIT for LIFE, Magazin für Fitness, Lauf- und Ausdauersport*, 10: 46–49.

Walking Pole Warriors Join Forces. 24.6.2003. http://walking.about.com/cs/poles/a/polepr0703.htm (accessed 29.10.2003).

Who is Nordic Walking? http://walking.about.com/library/weekly/aa071000f.htm (accessed 30.10.2002).

Company documents

Exel Plc Annual Reports 1997–2005.
Exel Plc Internet pages: www.exel.net (accessed 20.8.2002 and 13.5.2007).
Nordic Walking (2001) Video. Exel Plc.
Nordic Walking Internet pages: www.nordicwalking.com (accessed 13.5.2007).

The Spyder case

Secondary documents

"High techiä niché-alueilla. Spyderilla uusi suunta Suunnolle" (1997) *Arvopaperi* 1997, 5: 18–19.

Kotro, T. (2000) *Media and Mediators in the Product Development Process. Media Usage and the Transformations of Everyday Experience*, Turku: Media Studies, University of Turku. Medusa Research Project, Academy of Finland. www.uiah.fi/~tkotro/tanja_kotro_mediator.pdf (accessed 29.6.2004).

Kotro, T. (2001) "Lajituntemus jalostuu brandiksi," *Muoto*, 21, 2: 48–53.

Kotro, T. and Pantzar, M. (2002) "Product Development and Changing Cultural Landscapes – Is Our Future in 'Snowboarding?' ", *Design Issues*, 18, 2: 30–45.

"Lighted dive computer permits hands-free operation" (1998) *Design News*, 53, 15: 55, 57.

Pietiläinen, T. (2003) "Suunnon ja Polarin patenttiriita syveni," *Helsingin Sanomat*, 7.6.2003: D3.

Pirilä-Mänttäri, A. (2004) "Suunto on Eira vartalon kartalla," *Helsingin Sanomat*, 1.2.2004: E3.

Seeling, M. (1997) "Diving to success with Suunto diving computers," *Form Function Finland*, 65: 1, 24.

Company documents

Amer, Amer Group Plc's magazine for shareholders and customers.

Amer Sports Corporation Annual Reports 2000–2006.

Amer Sports Internet pages: www.amersports.com (accessed 19.7.2007).

Amer Sports, magazine for shareholders and customers.

Lähteenmäki, A. (2002) "Evolution of Suunto product lines/sports specific instruments," Internal material. Suunto Ltd. 20.5.2002.

Spyder, Instruction Manual. Suunto Ltd.

Suunto Ltd Internet pages: www.suunto.com (accessed 29.69.2004 and 14.6.2007).

Appendix 4
The most important publications utilized in the case selection

About SSH (2004) SSH Communications, www.ssh.com/company/about/history.html (accessed 15.12.2004).

The Benecol Story (2002) Nutra Ingredients.com, Europe, Breaking News on Supplements, Nutrition & Healthy Foods. www.nutraingredients.com/news/news-NG.asp?n=36575-the-benecol-story (accessed 21.10.2002).

Creative Thinking (2002) Handy Axes. Fiskars Corporation www.fiskars.com/our_company/12.html (accessed 5.7.2002).

Finnforest Kerto LVL (2002) Finnforest Corporation www.finnforest.com/product_list.asp?path=1;558;576;601;604 (accessed 8.7.2002).

Fiskars manufactures the world's first scissors with plastic handles (2002) Fiskars Corporation. www.fiskars.fi/corporation/h_1967.html (accessed 5.7.2002).

Flash Smelting (2004) Outokumpu, www.outokumpu.com/pages/Page____10010.aspx (accessed 29.10.2004).

GSM World – History of GSM (2002) www.gsmworld.com/about/history/history_page10.shtml (accessed 5.7.2002).

Helsinki implements Parkit parking payment service (2002) Press Release. Payway www.payway.fi/Press_info_e_020115.htm (accessed 15.1.2002).

Historia (2002) S. Pinomäki, www.spinomaki.fi/englanti/ehistoria.htm (accessed 9.7.2002).

Hyysalo, S. (2004) *Uses of Innovation. Wristcare in the practices of engineers and elderly*, Helsinki: University of Helsinki.

Ilvespää, H. (2003) Going for integration in high-quality papermaking. High Technology Finland 2003. www.hightechfinland.com/2003/newmaterialsprocess/forestindustry/metsoptoconcept.html (accessed 15.12.2004).

Interior Mudguards (2004) Leo Laine Oy, www.lokari.fi/eng_lokarit.html (accessed 10.11.2004).

Introduction (2002) Trafotek, www.trafotek.fi/en/en_tuotteet_yleista.php (accessed 8.7.2002).

Kervinen, J.-P. (2002) "Magneettinen muistimetalli on valmis valloituksiin," *Tekniikka & Talous*, 12.11.2002.

KONE Eco Disc (2002) Kone Corporation, www.kone.com/fi_FI/popup/0,,popup=yes&content=7607,00.html (accessed 4.7.2002).

"Lactose free milk – a new success story by Valio" (2002) *NORDICUM Scandinavian Business Magazine*, 2/2002: 66.

Laukkanen, R.M.T. and Virtanen P.K. (1998) "Heart Rate Monitors: State of the Art," *Journal of Sports Sciences*, 16, Supplement 1: 3–7.

Leino, R. (2001) "Marioff tyrmää tulipalot sammuttamalla," *Tekniikka & Talous*, 18.10.2001: 26.

"Lighted dive computer permits hands-free operation" (1998) *Design News* 53, 15: 55, 57.

Mackintosh, E., Mackintosh, I and Willmott, R. (1999) *Hybrid Mail in the Third Millenium*, United Kingdom: Mackintel Ltd, Kingsley.

Marinelog.com (2004) One million hours of Azipod operation, www.marinelog.com/DOCS/PRODS/MMIVprod0115a.html (accessed 7.12.2004).

Miettinen, R., Lehenkari, J., Hasu, M. and Hyvönen, J. (1999) *Osaaminen ja uuden luominen innovaatioverkoissa. Tutkimus kuudesta suomalaisesta innovaatiosta*, Helsinki: Sitra 226, Taloustieto Oy.

Mikkola, M. (2002) Keksintöesittelyssä: Abloy-lukko, astiankuivauskaappi, biohajoavat istukkeet, kolesterolia vähentävä margariini, Suomi-konepistooli www.yle.fi/teema/tiede/keksinnot/ (accessed 8.7.2002).

Milestones in R&D (2002) Valio Ltd, www.valio.fi/channels/konserni/eng/edellakavija/unnamed_1/unnamed.html (accessed 5.7.2002).

Multi-Source National Forest Inventory of Finland, History (2004) www.metla.fi/ohjelma/vmi/nfi-hist.htm (accessed 29.10 2004).

Nokia's first imaging phone marks start of Multimedia Messaging era (2001) Press Release, Nokia Mobile Phones <http://press.nokia.com/PR/200111/840889_5.html> (accessed 19.11.2001).

Nokia's Firsts in Telecommunications (2002) Nokia Corporation. www.nokia.com/about-nokia/inbrief/firsts.html (accessed 5.5.2002).

Paakki, J. and Kutvonen, P. (1999) "Mullistaako Linux ohjelmistoalan?" *Helsingin Sanomat*, 12.3.1999: D2.

Riihonen, R. (2002) "Vasen jalka edellä," *Kauppalehti Optio*, 16: 82–84.

Savustuspussi (2002) 2. palkinto Oy Hope Finlandia Ltd. Innosuomi. <http://innosuomi.iaf.fi/9499/1994.html> (accessed 9.7.2002).

Sensor Systems Division (2002) Vaisala, www.vaisala.fi/page.asp?Section=1651 (accessed 5.7.2002).

Siivonen, R. (2002) "Sauvakansan synty," *Ylioppilaslehti*, 89, 15: 10–11.

Steinbock, D. (2001) *The NOKIA Revolution. The Story of an Extraordinary Company that Transformed an Industry*, New York: AMACOM.

Vaisala History (2004) www.vaisala.com/page.asp?Section=8 (accessed 10.11.2004).

Veikkola, V. (1992) "Finnzymes tekee Suomen kalleimpia tuotteita," *Kauppalehti Optio*, 17: 51–52.

VTI History (2004) http://vti.fi/about/history.html (accessed 15.11.2004).

Yrittämisen juhlaa (1988) Helsinki: Yrittäjäin Fennia.

Öster, H. (1998) "Dekati mittaa hiukkaspäästöt sekunnin tarkkuudella," *Tekniikka & Talous*, 26.2.1998: 14–15.

Notes

1 Introduction

1 The word "customer" in this study includes both current and prospective customers. Thus, they may come from either unserved or served markets. This definition seems applicable here, since the customers for radical innovations often come from markets previously unserved by the firm.
2 A more detailed discussion on the concept follows in Section 2. The key concepts are defined and developed throughout the study. A glossary of the key concepts is given in Appendix 1.
3 For more information on the role of resources, see e.g. Penrose (1980) and Wernerfelt (1984).
4 Although recent studies (e.g. Ernst and Kim 2002) have accentuated the enabling role played by information technology in knowledge transfer, its role is likely to be more limited in the development of radical innovations, since knowledge is often created along the innovation development process, in which face-to-face communication plays an important role.
5 I.e. capabilities to recognize the value of new information, assimilate it and apply it to commercial ends (Cohen and Levinthal 1990: 128).
6 The very definitions have often been rather vague lists of attributes connected to proactive behavior (see e.g. Morgan 1992: 25–28).
7 The notion of context is important, since processes are affected by the context and, conversely, also shape it (Pettigrew 1997: 340–341).
8 According to O'Connor and McDermott (2004: 11), the development of a radical innovation typically takes ten years or more.
9 For further reading, see e.g. Zinkhan and Pereira (1994).
10 For a more detailed discussion on market-orientation research, see e.g. Lafferty and Hult (2001).
11 This idea is based on von Hippel's earlier studies (e.g. von Hippel 1986, 1989).

2 Proactiveness in the firm

1 Recently, the discussion on firm boundaries has been lively in numerous writings on virtual organizations (see e.g. Alexander 1997; Elliot 2006).
2 This kind of learning has also been called single-loop learning (Argyris 1994: 8–9; Hatch 1997: 372).
3 The rationality of organizational behavior has been widely questioned (e.g. Brunsson 2000; Cyert and March 1992; Mintzberg 1994). For a review of that discussion, see Stubbart (1989).
4 Chandler (1962) studied structural changes in firms. He concluded that the market is a determinant of business strategy and that strategy, in turn, defines the structure of the firm.

5 This contingency hypothesis was built on the foundations laid earlier by Burns and Stalker (1979; first edition in 1961).
6 For instance, see the emerged stream of studies on the fitness landscape (e.g. Beinhocker 1999; McKelvey 1999).
7 This is quite paradoxical, since, on the other hand, firms often cannot predict considerable changes, and all they can do is react to them (cf. Liuhto 1999).
8 Hamel and Prahalad (1994: 76) prefer using the word *foresight* instead of *vision* because according to them vision connotes a dream or an apparition, but industry foresight is based on a solid factual foundation.
9 First edition was published in 1969.
10 First edition was published in 1959.
11 First edition was published in 1963.
12 This limited search has its advantages since it is often cost-effective, reduces uncertainty, and furthermore, the existence of patterns in a firm's behavior signifies that there is a certain consistency over time (Araujo and Easton 1996: 368; Miles and Snow 1978: 157). However, if managers do not periodically impugn the behavior patterns, then there is a danger that the potential costs of limited search may become too high (Miles and Snow 1978: 157).
13 For further discussion on the concept of adaptation, see e.g. Hrebiniak and Joyce (1985: 337) and Miles and Snow (1978: 3).
14 For further discussion on the concept of flexibility, see e.g. Bahrami (1992: 36), Brunsson (1989: 221) and Volberda and Rutges (1999: 101).

3 Proactiveness toward customers

1 This definition reflects a behavioral orientation and thus is in line with the concepts of proactive and reactive *behavior*. However, market orientation could be also defined as a culture (see e.g. Day 1994: 43).
2 Examples of exogenous market factors are government regulations, competitor actions and other environmental forces (Kohli and Jaworski 1990: 3).
3 Closely related concepts include "sales-driven" and "customer-driven": for a more detailed conceptual discussion, see Kumar *et al.* (2000: 132).
4 This kind of behavior has also been called market driving (Kumar *et al.* 2000; cf. Hills and Sarin 2003: 17).
5 A variety of specific need-assessment techniques is presented by Holt *et al.* (1984: 56–121), for example.
6 Hamel and Prahalad (1994: 100) characterized it as behavior in which firms lead customers where they want to go but do not know it yet, i.e. firms "amaze" their customers.
7 As stated earlier, market orientation consists of both market proactiveness and market reactiveness (Jaworski *et al.* 2000; Kohli and Jaworski 1990: 3).
8 See e.g. Delene *et al.* (1997); Solberg (2002); Szymanski *et al.* (1993).
9 See e.g. Delene *et al.* (1997); Lemak and Arunthanes (1997); Sheth and Parvatiyar (2001); Solberg (2002); Szymanski *et al.* (1993) and Zou and Cavusgil (1996).
10 First edition was published in 1963.
11 According to Hamel (2000: 40–43), shareholder returns increased considerably in the late 1990s, which diminished the sphere of strategic operations in many firms.
12 The terms "collaboration" and "cooperation" are used interchangeably throughout the text.

4 Developing radical innovations

1 For more information on dissonance reduction, see e.g. Festinger (1964).
2 Disruptive technologies are technologies that bring to the market a very different

value proposition than had been available previously and create entirely new markets and business models (Christensen 1997: xv; Christensen *et al.* 2002). Thus, they tend to create the most radical of radical innovations.

3 According to the study conducted by Mullins and Sutherland (1998: 228), managers' knowledge of customers and their needs is sharply limited in these markets. In the early stages of innovation development in particular, there is often nothing more than a rather vague feeling for a possible market (cf. Millier 1999: 50–51).

4 The term "development stage" is used in this study to refer to the particular stage, whereas "development process" and "development" refer to the whole process of which the development stage is only a part.

5 This approach was advocated by Rogers (1983: 157), for example, who claimed that innovation tracer studies were often too limited because they included only part of the process, i.e. they typically ended at the commercialization phase.

6 Nevertheless, the market has traditionally been seen as one of the major sources of incremental innovation. After all, it is here that firms eventually realize their profits (Urban and Hauser 1993: 120).

7 I.e. direct discussions between customers and product developers (Deszca *et al.* 1999: 624).

8 I.e. gathering, analyzing and applying information originating from observations in the field (Leonard and Rayport 1997: 104).

9 For an extensive discussion on first-mover advantages that a fast development process can offer, see Lieberman and Montgomery (1988). See also the discussion on first-mover disadvantages in Tellis and Golder (1996) and Lieberman and Montgomery (1998).

10 Defining the product too explicitly and too early may prevent responsiveness later (Khurana and Rosenthal 1998: 65–66).

11 Due to the focus on radical innovations and customers in this study, this measure seems to be adequate. In general, the selection of appropriate measures of new-product performance is difficult. Griffin and Page (1993), for example, found 75 different measures in the literature.

12 Innovative adopters consist of innovators and early adopters (Easingwood and Koustelos 2000: 29; Rogers 1983: 247). See Section 4.6.3.

13 Innovative adopters are also referred to as the "early market" (Moore 1999b: 16). Although they may often have similar characteristics as lead users, there is an important difference between the two groups: lead users participate in the innovation development and create solutions, whereas innovative adopters adopt commercial solutions (cf. von Hippel *et al.* 2000: 23).

14 Also called pragmatists (Moore 1999a: 40).

5 Customer-related proactiveness in the case innovations

1 This article concentrated especially on market proactiveness at the launching of radical innovations.

2 This article focused on the manifestation of the international context in market proactiveness during the radical innovation development process.

3 This article provided an overview of the whole case study, including the organizational context and environmental factors.

4 The most important secondary references utilized in the case-study selection are given in Appendix 4. The refereed discussions and e-mail correspondence are listed in Appendix 2.

5 The original citations were in Finnish and were translated later into English. In order to minimize the researcher's influence, the translations were done by an outsider, who was unfamiliar with the theoretical framework and with the secondary data used in the case descriptions.

6 The thermostability of a polymerase is usually characterized by its half-life at a specific temperature.

7 The 3' → 5' exonuclease activity allows the polymerase to correct a mistake if it incorporates an incorrect nucleotide. When it is found that the base incorporated was incorrect, it removes the incorrect base-paired nucleotide so that the polymerase can try again to get it right.

8 In fact, one kilogram would saturate the world market for the next 2000 years.

9 I.e. transposable elements of DNA, called "jumping genes."

10 Around 90 per cent of letters in Finland are sent by organizations (Nikali 1998: 22).

11 In this study, the terms "Finland Post" and "Post" refer to the whole Finland Post Group, currently known as the Itella Group. The Finland Post Corporation became the Itella Corporation, and the Finland Post Group became the Itella Group in June 2007. New names were adopted because of the Group's more diversified and internationalized business operations.

12 "ePost Mail" was the term used in the international communication. However, since the Finland Post used the term "ePost Letter" in its own English-language communication, the latter is used in this study.

13 At first, all the centers had the same data-handling equipment, but by 1997, given the more advanced data-processing equipment, all the data receiving was concentrated in Helsinki, and other centers became local offices that could only print, envelope and deliver ePost Letters.

14 It was particularly difficult to compete with the software houses, e.g. Tieto-Savo (later Enfo) and Tietotehdas (later TietoEnator), that also provided printing services for their customers. Since the printing service was only the by-product of the software that they were selling, they were able to sell it at a low price.

15 Hi-Fog fire-extinguishing technology is protected by more than 600 patents or patent applications all around the world.

16 This resembles the situation that is characterized by Allen (2003: 19) as follows: "Sometimes a company is formed around a creative genius who carries the vision for the company and without whom the company would have no purpose."

17 Source: Puustinen, Terho (2001) Hyvin sammutettu. [Well extinguished]. *Talouselämä* 2.12.2001, 24–26.

18 It is the United Nations' specialized agency that is responsible for improving maritime safety and preventing pollution.

19 *MS Frans Suell* (later *MS Gabriella*) and *MS Crown* of Scandinavia.

20 The time lag between the order and delivery (i.e. final payment) may have been several years if the vessel was under construction.

21 E.g. whereas the first sprinklers in 1991 protected an area of 4m^2, the one launched in 2000 covers 25m^2.

22 NFPA is a US-based international organization that aims to develop international safety codes and standards for fire prevention.

23 VdS Loss Prevention is an institute owned by German insurers.

24 This reasoning is supported by e.g. Spencer (2003: 221–222), who states that competing firms with a joint interest in an emerging technology may together influence their institutional environment so that it supports the new technology.

25 Power walking was launched in the United States in the early 1990s, but it has never really caught on.

26 However, these studies on basic trait correlates of athleticism should be treated with caution, since it has also been argued that research findings on athletes' traits are inconsistent and that there is no distinguishable "athletic personality" (Vealey 1992).

27 By the end of 1998, The Central Association for Recreational Sports and Outdoor Activities had trained more than 2,000 instructors around Finland.

28 Cooperation agreements have been made with famous athletes such as Bjørn Dæhlie and Stephan Eberharter.

29 It has been argued that in the context of innovations, the imagination used in fusing earlier inventions and ideas may be even more important than the originality (Evans 2004: 25).

30 It has been argued that, in practice, many firms in the consumer-goods industry tend to use their employees for customer-need-assessment purposes during the innovation development process (Holt *et al.* 1984: 35–36).

31 However, it should be acknowledged that there was also slight hesitation among some employees that this concept might cannibalize Suunto's own diving-only instrument sales. This anxiety was later proven to be unfounded since in many cases customers bought both a diving-only computer and a Spyder.

32 As stated in the previous section, the development project was running late due to problems related to the software development, thus the testing period was shorter than planned.

6 A modified framework of customer-related proactiveness during the development process

1 The role of serendipity has also been emphasized in the various radical innovation development processes described by Nayak and Ketteringham (1986).

2 All in all, consumers facing a high cost of product failure are slow in adopting new innovations (Porter 1980: 227).

7 Conclusions

1 This is in line with Golden (1992: 855), who argued that retrospective memories of past accounts and behaviors tend to be more accurate than memories of past beliefs and intentions.

2 A step in this direction is taken in Sandberg (2007b).

Bibliography

English-speaking readers should note that the alphabetical order is sometimes affected by the fact that in Scandinavia accented vowels are alphabetized last, e.g. å, ä and ö come after z. Thus, Kumar precedes Kärkkäinen, for instance.

Aaker, D.A. and Mascarenhas, B. (1984) "The Need for Strategic Flexibility," *Journal of Business Strategy*, 5, 2: 74–82.

Abell, D.F. (1978) "Strategic Windows," *Journal of Marketing*, 42, 3: 21–26.

—— (1999) "Competing Today While Preparing for Tomorrow," *MIT Sloan Management Review*, 40, 3: 73–81.

Acs, Z.J., Morck, R.K. and Yeung, B. (2001) "Entrepreneurship, Globalization, and Public Policy," *Journal of International Management*, 7, 3: 235–251.

Adams, M.E., Day, G.S. and Dougherty, D. (1998) "Enhancing New Product Development Performance: An Organizational Learning Perspective," *Journal of Product Innovation Management*, 15, 5: 403–422.

Adner, R. and Levinthal, D.A. (2002) "The Emergence of Emerging Technologies," *California Management Review*, 45, 1: 50–66.

Ahmed, P.K. (1998) "Culture and Climate for Innovation," *European Journal of Innovation Management*, 1, 1: 30–43.

Alexander, M. (1997) "Getting to Grips with the Virtual Organization," *Long Range Planning*, 30, 1: 122–124.

Ali, A. (1994) "Pioneering Versus Incremental Innovation: Review and Research Propositions," *Journal of Product Innovation Management*, 11, 1: 46–61.

Allen, K.R. (2003) *Bringing New Technology to Market*, Upper Saddle River: Prentice Hall.

Alvesson, M. and Berg, P.O. (1992) *Corporate Culture and Organizational Symbolism*, Berlin: Walter de Gruyter.

Anderson, P. and Tushman, M.L. (1990) "Technological Discontinuities and Dominant Designs: A Cyclical Model of Technological Change," *Administrative Science Quarterly*, 35, 4: 604–633.

Ansoff, I. (1958) "A Model for Diversification," *Management Science*, 4, 4: 392–414.

Aragon-Correa, J.A. (1998) "Strategic Proactivity and Firm Approach to the Natural Environment," *Academy of Management Journal*, 41, 5: 556–567.

Araujo, L. and Easton, G. (1996) "Strategy: Where is the Pattern?," *Organization*, 3, 3: 361–383.

Arbnor, I. and Bjerke, B. (1997) *Methodology for Creating Business Knowledge*, Thousand Oaks: Sage Publications.

Argyris, C. (1994) *On Organizational Learning*, Cambridge: Blackwell Publishers.

Ashby, W.R. (1971) *An Introduction to Cybernetics* (1st published in 1956), London: Chapman & Hall.

Ashley, W.C. and Morrison, J.L. (1997) "Anticipatory Management: Tools for Better Decision Making," *The Futurist*, 31, 5: 47–50.

Astley, W.G. and Van de Ven, A.H. (1983) "Central Perspectives and Debates in Organization Theory," *Administrative Science Quarterly*, 28, 2: 245–273.

Atuahene-Gima, K. (1996) "Market Orientation and Innovation," *Journal of Business Research*, 35, 2: 93–103.

Atuahene-Gima, K. and Evangelista, F. (2000) "Cross-Functional Influence in New Product Development: An Exploratory Study of Marketing and R&D Perspectives," *Management Science*, 46, 10: 1269–1284.

Audretsch, D.B. and Feldman, M.P. (1996) "R&D Spillovers and the Geography of Innovation and Production," *The American Economic Review*, 86, 3: 630–640.

Avlonitis, G.J., Papastathopoulou, P.G. and Gounaris, S.P. (2001) "An Empirically-Based Typology of Product Innovativeness for New Financial Services: Success and failure scenarios," *Journal of Product Innovation Management*, 18, 5: 324–342.

Ayal, I. and Zif, J. (1979) "Market Expansion Strategies in Multinational Marketing," *Journal of Marketing*, 43, 2: 84–94.

Bacon, G., Beckman, S., Mowery, D. and Wilson, E. (1994) "Managing Product Definition in High-Technology Industries: A Pilot Study," *California Management Review*, 36, 3: 32–56.

Bahrami, H. (1992) "The Emerging Flexible Organization: Perspectives from Silicon Valley," *California Management Review*, 34, 4: 33–51.

Baker, M. and Hart, S. (1999) *Product Strategy and Management* (1st published in 1988), Essex: Prentice Hall.

Barnett, C.K. and Pratt, M.G. (2000) "From Threat-Rigidity to Flexibility. Toward a Learning Model of Autogenic Crisis in Organizations," *Journal of Organizational Change Management*, 13, 1: 74–88.

Barr, P.S. (1998) "Adapting to Unfamiliar Environmental Events: A Look at the Evolution of Interpretation and Its Role in Strategic Change," *Organization Science*, 9, 6: 644–669.

Barrett, H. and Weinstein, A. (1998) "The Effect of Market Orientation and Organizational Flexibility on Corporate Entrepreneurship," *Entrepreneurship: Theory and Practice*, 23, 1: 57–70.

Bateman, T.S. and Crant, J.M. (1993) "The Proactive Component of Organizational Behavior: A Measure and Correlates," *Journal of Organizational Behavior*, 14, 2: 103–118.

—— (1999) "Proactive Behavior: Meaning, Impact, Recommendations," *Business Horizons*, 42, 3: 63–70.

Beard, C. and Easingwood, C. (1996) "New Product Launch. Marketing Action and Launch Tactics for High-Technology Products," *Industrial Marketing Management*, 25, 2: 87–103.

Beinhocker, E.D. (1999) "Robust adaptive strategies," *MIT Sloan Management Review*, 40, 3: 95–106.

Beise, M. (2004) "Lead Markets: Country-Specific Drivers of the Global Diffusion of Innovations," *Research Policy*, 33, 6–7: 997–1018.

Bem, D.J. and Allen, A. (1974) "On Predicting Some of the People Some of the Time: The Search for Cross-Situational Consistencies in Behavior," *Psychological Review*, 81, 6: 506–520.

Benfari, R.C., Wilkinson, H.E. and Orth, C.D. (1986) "The Effective Use of Power," *Business Horizons*, 29, 3: 12–16.

Bennett, R. (1996) *Corporate Strategy and Business Planning*, London: Pitman Publishing.

Berthon, P., Hulbert, J.M. and Pitt, L. (2004) "Innovation or Customer Orientation? An Empirical Investigation," *European Journal of Marketing*, 38, 9–10: 1065–1090.

Bohn, R. (2000) "Stop Fighting Fires," *Harvard Business Review*, 78, 4: 82–91.

Boisot, M.H. (1994) "Learning as Creative Destruction: The Challenge for Eastern Europe," in R. Boot, J. Lawrence and J. Morris (eds) *Managing the Unknown by Creating New Futures*, London: McGraw-Hill, pp. 41–56.

Boisot, M. and Child, J. (1999) "Organizations as Adaptive Systems in Complex Environments: The Case of China," *Organization Science*, 10, 3: 237–252.

Bourgeois, L.J. III (1981) "On the Measurement of Organizational Slack," *Academy of Management Review*, 6, 1: 29–39.

—— (1984) "Strategic Management and Determinism," *Academy of Management Review*, 9, 4: 586–596.

Brown, A. (1992) "Organizational Culture: The Key to Effective Leadership and Organizational Development," *Leadership & Organization Development Journal*, 13, 2: 3–6.

Brown, T.J., Mowen, J.C., Donavan, D.T. and Licata, J.W. (2002) "The Customer Orientation of Service Workers: Personality Trait Effects on Self- and Supervisor Performance Ratings," *Journal of Marketing Research*, 39, 1: 110–119.

Brownlie, D. and Spender, J.C. (1995) "Managerial Judgement in Strategic Marketing: Some Preliminary Thoughts," *Management Decision*, 33, 6: 39–50.

Bruno, A.V. (1989) "Marketing Lessons from Silicon Valley for Technology-Based Firms," in R.W. Smilor (ed.) *Customer-Driven Marketing. Lessons from Entrepreneurial Technology Companies*, Lexington: Lexington Books, pp. 33–44.

Brunsson, N. (1989) "Administrative Reforms as Routines," *Scandinavian Journal of Management*, 5, 5: 219–228.

—— (2000) *The Irrational Organization. Irrationality as a Basis for Organizational Action and Change* (1st published in 1985 by John Wiley & Sons Ltd), Bergen: Fagbokforlaget Vigmostad & Bjørke.

Buganza, T., Dell'era, C. and Verganti, R. (2004) "Exploring the Relationships between Product Development and Environmental Turbulence: The Case of Mobile TLC Services," *Proceedings of the 11th International Product Development Management Conference*, Dublin, Ireland.

Burgelman, R.A. and Maidique, M.A. (1988) *Strategic Management of Technology and Innovation*, Homewood: Irwin.

Burns, T. (1963) "Industry in a New Age," *New Society*, 31, 1: 17–20.

Burns, T. and Stalker, G.M. (1979) *The Management of Innovation* (1st published in 1961), London: Tavistock Publications.

Burrell, G. and Morgan, G. (1988) *Sociological Paradigm and Organisational Analysis* (1st published in 1979 by Heinemann Educational Books), Hants: Gower Publishing Company.

Buss, D.M. (1991) "Evolutionary Personality Psychology," *Annual Review of Psychology*, 42: 459–491.

Calori, R. and Sarnin, P. (1991) "Corporate Culture and Economic Performance: A French Study," *Organization Studies*, 12, 1: 49–74.

Camillus, J.C. and Datta, D.K. (1991) "Managing Strategic Issues in a Turbulent Environment," *Long Range Planning*, 24, 2: 67–74.

Campbell, D.J. (2000) "The Proactive Employee: Managing Workplace Initiative," *Academy of Management Executive*, 14, 3: 52–66.

Carson, D., Gilmore, A., Perry, C. and Gronhaug, K. (2001) *Qualitative Marketing Research*, London: Sage Publications.

Chaganti, R. and Sambharya, R. (1987) "Strategic Orientation and Characteristics of Upper Management," *Strategic Management Journal*, 8, 4: 393–401.

Chakravarthy, B.S. (1982) "Adaptation: A Promising Metaphor for Strategic Management," *Academy of Management Review*, 7, 1: 35–44.

—— (1986) "Measuring Strategic Performance," *Strategic Management Journal*, 7, 5: 437–458.

Chandler, A.D. Jr. (1962) *Strategy and Structure: Chapters in the History of Industrial Enterprise*, Cambridge: MIT Press.

Chandy, R.K. and Tellis, G.J. (1998) "Organizing for Radical Product Innovation: The Overlooked Role of Willingness to Cannibalize," *Journal of Marketing Research*, 35, 4: 474–487.

—— (2000) "The Incumbent's Curse? Incumbency, Size, and Radical Product Innovation," *Journal of Marketing*, 64, 3: 1–17.

Chaney, P.K., Devinney, T.M. and Winer, R.S. (1991) "The Impact of New Product Introductions on the Market Value of Firms," *Journal of Business*, 64, 4: 573–610.

Chattopadhyay, P., Glick, W.H. and Huber, G.P. (2001) "Organizational Actions in Response to Threats and Opportunities," *Academy of Management Journal*, 44, 5: 937–955.

Child, J. (1972) "Organizational Structure, Environment and Performance: The Role of Strategic Choice," *Sociology*, 6, 1: 1–22.

Christensen, C.M. (1997) *The Innovator's Dilemma. When New Technologies Cause Great Firms to Fail*, Boston: Harvard Business School Press.

Christensen, C.M., Johnson, M.W. and Rigby, D.K. (2002) "Foundations for Growth. How To Identify and Build Disruptive New Businesses," *MIT Sloan Management Review*, 43, 3: 22–31.

Chryssochoidis, G.M. and Wong, V. (1996) "Rolling Out New Products Across International Markets – Causes of Delays," *Proceedings of the 3rd International Product Development Management Conference*, Fontainebleau, France.

Church, T.S., Earnest, C.P. and Morss, G.M. (2002) 'Field Testing of Physiological Responses Associated with Nordic Walking', *Research Quarterly for Exercise and Sport*, 73, 3: 296–300.

Cohen, W.M. and Levinthal, D.A. (1990) "Absorptive Capacity: A New Perspective on Learning and Innovation," *Administrative Science Quarterly*, 35, 1: 128–152.

Collins, M. and Bloom, R. (1991) "The Role of Oral History in Accounting," *Accounting, Auditing & Accountability Journal*, 4, 4: 23–31.

Cook, K., Shortell, S.M., Conrad, D.A. and Morrisey, M.A. (1983) "A Theory of Organizational Response to Regulation: The case of Hospitals," *Academy of Management Review*, 8, 2: 193–205.

Cooper, L.G. (2000) "Strategic Marketing Planning for Radically New Products," *Journal of Marketing*, 64, 1: 1–16.

Cooper, R.G. (1979) "The Dimensions of Industrial New Product Success and Failure," *Journal of Marketing*, 43, 3: 93–103.

—— (1988) "The New Product Process: A Decision Guide for Management," *Journal of Marketing Management*, 3, 3: 238–255.

—— (2000) *Product Leadership. Creating and Launching Superior New Products*, Cambridge: Perseus Books.

Courneya, K.S. and Hellsten, L.-A.M. (1998) "Personality Correlates of Exercise Behavior, Motives, Barriers and Preferences: An Application of the Five-Factor Model," *Personality and Individual Differences*, 24, 5: 625–633.

Cova, B., Mazet, F. and Salle, R. (1994) "From Competitive Tendering to Strategic Marketing: An Inductive Approach for Theory-building," *Journal of Strategic Marketing*, 2, 1: 29–47.

Covin, J.G and Slevin, D.P. (1988) "The Influence of Organization Structure on the Utility of an Entrepreneurial Top Management Style," *Journal of Management Studies*, 25, 3: 217–234.

Crawford, C.M. (1977) "Marketing Research and the New Product Failure Rate," *Journal of Marketing*, 41, 2: 51–61.

—— (1997) *New Products Management* (5th edn, 1st edn in 1983), Chicago: Irwin.

Culkin, N., Smith, D. and Fletcher, J. (1999) "Meeting the Information Needs of Marketing in the Twenty-First Century," *Marketing Intelligence & Planning*, 17, 1: 6–12.

Cumming, B.S. (1998) "Innovation Overview and Future Challenges," *European Journal of Innovation Management*, 1, 1: 21–29.

Cyert, R.M. and March, J.G. (1992) *A Behavioral Theory of the Firm* (2nd edn, 1st edn in 1963 by Prentice-Hall), Cambridge: Blackwell.

Dandridge, T.C., Mitroff, I. and Joyce, W.F. (1980) "Organizational Symbolism: A Topic to Expand Organizational Analysis," *Academy of Management Review*, 5, 1: 77–82.

Darling, J.R. and Nurmi, R. (1996) "Downsizing the Finnish Company: A Call for Managerial Leadership," in A. Suominen (ed.) *Johtaminen murroksessa. Management in Transition*, Turku: Turku School of Economics and Business Administration, pp. 251–272.

D'Aveni, R.A. (1999) "Strategic Supremacy through Disruption and Dominance," *MIT Sloan Management Review*, 40, 3: 127–135.

D'Aveni, R.A. and MacMillan, I.C. (1990) "Crisis and the Content of Managerial Communications: A Study of the Focus of Attention of Top Managers in Surviving and Failing Firms," *Administrative Science Quarterly*, 35, 4: 634–657.

Day, G.S. (1994) "The Capabilities of Market-Driven Organizations," *Journal of Marketing*, 58, 4: 37–52.

Day, G.S. and Wensley, R. (1988) "Assessing Advantage: A Framework for Diagnosing Competitive Superiority," *Journal of Marketing*, 52, 2: 1–20.

de Brentani, U. (2001) "Innovative Versus Incremental New Business Services: Different Keys for Achieving Success," *Journal of Product Innovation Management*, 18, 3: 169–187.

de Geus, A.P. (1988) "Planning as Learning," *Harvard Business Review*, 66, 2: 70–74.

de Heer, J., Groenland, E.A.G. and Schoormand, J.P.L. (2002) *How to Obtain Consumer Requirements for the Initial Conceptualization of New Products: Toward a Context-Aware Consumer Latent Needs Methodology*, Enschede: Telematica Instituut Corporate Report Series.

de Moerloose, C. (2000) "New Product Success Factors, New Product Typology and Moderating Impact on Success Factors According to the Type of Newness," *Proceedings of the 7th International Product Development Management Conference*, Leuven, Belgium.

Dekimpe, M.G., Parker, P.M. and Sarvary, M. (2000) "Global Diffusion of Technological Innovations: A Coupled-Hazard Approach," *Journal of Marketing Research*, 37, 1: 47–59.

Delene, L.M., Meloche, M.S. and Hodskins, J.S. (1997) "International Product Strategy:

Building the Standardisation-Modification Decision," *Irish Marketing Review*, 10, 1: 47–54.

Deszca, G., Munro, H. and Noori, H. (1999) "Developing Breakthrough Products: Challenges and Options for Market Assessment," *Journal of Operations Management*, 17, 6: 613–630.

Devinney, T.M. (1995) "Significant Issues for the Future of Product Innovation," *Journal of Product Innovation Management*, 12, 1: 70–75.

Di Benedetto, C.A. (1999) "Identifying the Key Success Factors in New Product Launch," *Journal of Product Innovation Management*, 16, 6: 530–544.

Dill, W.R. (1958) "Environment as an Influence on Managerial Autonomy," *Administrative Science Quarterly*, 2, 4: 409–443.

Donaldson, L. (1997) "A Positivist Alternative to the Structure-Action Approach," *Organization Studies*, 18, 1: 77–92.

Dougherty, D. (1996) "Organizing for Innovation.," in S.R. Clegg, C. Hardy and W.R. Nord (eds) *Handbook of Organization Studies*, London: Sage Publications, pp. 424–439.

Douglas, S.P. and Craig, C.S. (1989) "Evolution of Global Marketing Strategy: Scale, Scope and Synergy," *Columbia Journal of World Business*, 24, 3: 47–59.

Doz, Y.L and Prahalad, C.K. (1991) "Managing DMNCs: A Search for a New Paradigm," *Strategic Management Journal*, 12, Special Issue: 145–164.

Drucker, P.F. (1974) *Management: Tasks, Responsibilities, Practices*, London: Heinemann.

—— (1985) "The Discipline of Innovation," *Harvard Business Review*, 63, 3: 67–72.

—— (1997) "The Future that has Already Happened," *Harvard Business Review*, 75, 5: 20–24.

Dunn, D.T. Jr. and Thomas, C.A. (1986) "Strategy for Systems Sellers: A Grid Approach," *Journal of Personal Selling & Sales Management*, 6, 2: 1–10.

Dunphy, S. and Herbig, P.A. (1995) "Acceptance of Innovations: The Customer is the Key!," *The Journal of High Technology Management Research*, 6, 2: 193–209.

Dvir, D., Segev, E. and Shenhar, A. (1993) "Technology's Varying Impact on the Success of Strategic Business Units within the Miles and Snow Typology," *Strategic Management Journal*, 14, 2: 155–162.

Easingwood, C. and Koustelos, A. (2000) "Marketing High Technology: Preparation, Targeting, Positioning, Execution," *Business Horizons*, 43, 3: 27–34.

Eisenhardt, K.M. (1989) "Building Theories from Case Study Research," *Academy of Management Review*, 14, 4: 532–550.

Eisenhardt, K.M. and Tabrizi, B.N. (1995) "Accelerating Adaptive Processes: Product Innovation in the Global Computer Industry," *Administrative Science Quarterly*, 40, 1: 84–110.

El Louadi, M. (1998) "The Relationship Among Organization Structure, Information Technology and Information Processing in Small Canadian Firms," *Canadian Journal of Administrative Sciences*, 15, 2: 180–199.

Elliot, S. (2006) "Technology-Enabled Innovation, Industry Transformation and the Emergence of Ambient Organizations," *Industry and Innovation*, 13, 2: 209–225.

Emerson, R.M. (1962) "Power-Dependence Relations," *American Sociological Review*, 27, 1: 31–41.

Ernst, D. and Kim, L. (2002) "Global Production Networks, Knowledge Diffusion, and Local Capability Formation," *Research Policy*, 31, 8–9: 1417–1429.

Eskola, J. and Suoranta, J. (1998) *Johdatus laadulliseen tutkimukseen*, Tampere: Vastapaino.

Ettlie, J.E. (1997) "Integrated Design and New Product Success," *Journal of Operations Management*, 15, 1: 33–55.

Evans, H. (2004) "What Drives America's Great Innovators?" *Fortune*, October 1, 2004: 24–25.

Evans, J.S. (1991) "Strategic Flexibility for High Technology Manoeuvres: A Conceptual Framework," *Journal of Management Studies*, 28, 1: 68–89.

Festinger, L. (1964) "Behavioral Support for Opinion Change," *Public Opinion Quarterly*, 28, 3: 404–417.

Floyd, S.W. and Lane, P.J. (2000) "Strategizing Throughout the Organization: Management Role Conflict in Strategic Renewal," *The Academy of Management Review*, 25, 1: 154–177.

Forster, N. (1994) "The Analysis of Company Documentation," in C. Cassell and G. Symon (eds) *Qualitative Methods in Organizational Research: A Practical Guide*, London: Sage Publications, pp. 147–166.

Forte, M., Hoffman, J.J., Lamont, B.T. and Brockmann, E.N. (2000) "Organizational Form and Environment: An Analysis of Between-Form and Within-Form Responses to Environmental Change," *Strategic Management Journal*, 21, 7: 753–773.

Gardner, D.M., Johnson, F., Lee, M. and Wilkinson, I. (2000) "A Contingency Approach to Marketing High Technology Products," *European Journal of Marketing*, 34, 9–10: 1053–1077.

Gassmann, O. and von Zedtwitz, M. (1999) "New Concepts and Trends in International R&D Organization," *Research Policy*, 28, 2–3: 231–250.

Gatignon, H. and Robertson, T.S. (1985) "A Propositional Inventory for New Diffusion Research," *Journal of Consumer Research*, 11, 4: 849–867.

Gemünden, H.G., Ritter, T. and Heydebreck, P. (1996) "Network Configuration and Innovation Success: An Empirical Analysis in German High-tech Industries," *International Journal of Research in Marketing*, 13, 5: 449–462.

Gemünden, H.G., Salomo, S. and Krieger, A. (2005) "The Influence of Project Autonomy on Project Success," *International Journal of Project Management*, 23, 5: 366–373.

Gertler, M.S. (1995) "Being There: Proximity, Organization, and Culture in the Development and Adoption of Advanced Manufacturing Technologies," *Economic Geography*, 71, 1: 1–26.

Gerybadze, A. and Reger, G. (1999) "Globalization of R&D: Recent Changes in the Management of Innovation in Transnational Corporations," *Research Policy*, 28, 2–3: 251–274.

Ghauri, P. (2004) "Designing and Conducting Case Studies in International Business Research," in R. Marschan-Piekkari and C. Welch (eds) *Handbook of Qualitative Research Methods for International Business*, Cheltenham: Edward Elgar, pp. 109–124.

Ghauri, P. and Grønhaug, K. (2002) *Research Methods in Business Studies. A Practical Guide* (2nd edn, 1st edn in 1995), Harlow: Pearson Education.

Ginsberg, A. and Venkatraman, N. (1985) "Contingency Perspectives of Organizational Strategy: A Critical Review of the Empirical Research," *Academy of Management Review*, 10, 3: 421–434.

Glaser, B.R. and Strauss, A.L. (1967) *The Discovery of Grounded Theory: Strategies for Qualitative Research*, New York: Aldine de Gruyter.

Gobeli, D.H. and Brown, D.J. (1993) "Improving the Process of Product Innovation," *Research Technology Management*, 36, 2: 38–44.

Goffin, K. (1998) "Evaluating Customer Support During New Product Development – An Exploratory Study," *Journal of Product Innovation Management*, 15, 1: 42–56.

Golden, B.R. (1992) "The Past Is the Past – or Is It? The Use of Retrospective Accounts as Indicators of Past Strategy," *Academy of Management Journal*, 35, 4: 848–860.

Golder, P.N. (2000a) "Historical Method in Marketing Research with New Evidence on Long-Term Market Share Stability," *Journal of Marketing Research*, 37, 2: 156–172.

—— (2000b) "Insights from Senior Executives about Innovation in International Markets," *Journal of Product Innovation Management*, 17, 5: 326–340.

Golder, P.N. and Tellis, G.J. (1997) "Will It Ever Fly? Modeling the Takeoff of Really New Consumer Durables," *Marketing Science*, 16, 3: 256–270.

Griffin, A. and Page, A.L. (1993) "An Interim Report on Measuring Product Development Success and Failure," *Journal of Product Innovation Management*, 10, 4: 291–308.

Grönroos, C. (2000) *Service Management and Marketing. A Customer Relationship Management Approach* (2nd edn), Chichester: John Wiley & Sons.

Guiltinan, J.P. (1999) "Launch Strategy, Launch Tactics, and Demand Outcomes," *Journal of Product Innovation Management*, 16, 6: 509–529.

Gummesson, E. (1991) *Qualitative Methods in Management Research*, Newbury Park: Sage Publications.

Gupta, A.K. and Wilemon, D.L. (1990) "Accelerating the Development of Technology-Based New Products," *California Management Review*, 32, 2: 24–44.

Gupta, A.K., Raj, S.P. and Wilemon, D. (1986) "A Model for Studying R&D – Marketing Interface in the Product Innovation Process," *Journal of Marketing*, 50, 2: 7–17.

Halinen, A. (1994) *Exchange Relationships in Professional Services. A Study of Relationship Development in the Advertising Sector*, Turku: Turku School of Economics and Business Administration.

Halinen, A. and Törnroos, J.-Å. (1995) "The Meaning of Time in the Study of Industrial Buyer–Seller Relationships," in K. Möller and D. Wilson (eds) *Business Marketing: An Interaction and Network Perspective*, Boston: Kluwer Academic Publishers, pp. 493–529.

Hambrick, D.C. (1983) "Some Tests of the Effectiveness and Functional Attributes of Miles and Snow's Strategic Types," *Academy of Management Journal*, 26, 1: 5–26.

Hamel, G. (2000) *Leading the Revolution*, Boston: Harvard Business School Press.

Hamel, G. and Prahalad, C.K. (1989) "Strategic Intent," *Harvard Business Review*, 67, 3: 63–76.

—— (1994) *Competing for the future*, Boston: Harvard Business School Press.

Handy, C. (1997) "Finding Sense in Uncertainty," in Rowan Gibson (ed.) *Rethinking the Future. Rethinking Business, Principles, Competition, Control & Complexity*, London: Nicholas Brealey Publishing, pp. 16–33.

Hansén, S.-O. (1981) *Studies in Internationalisation of the Pharmaceuticals Industry – A Taxonomic Approach*, Turku: Research Institute of the Åbo Akademi Foundation.

—— (1990) "Luovuuden esiinsaaminen, johtajuuden avaintekijä," *Futura*, 3, 3: 55–65.

—— (1991) The Two Contexts; Conceptualization and Realization," in Juha Näsi (ed.) *Arenas of Strategic Thinking*, Helsinki: Foundation for Economic Education, pp. 127–132.

Hansén, S.-O. and Wakonen, J. (1997) "Innovation, A Winning Solution?" *International Journal of Technology Management*, 13, 4: 345–358.

Harper, S.C. (2000) "Timing – The Bedrock of Anticipatory Management," *Business Horizons*, 43, 1: 75–83.

Harris, L.C. and Piercy, N.F. (1997) "Market Orientation is Free: The Real Costs of Becoming Market-Led," *Management Decision*, 35, 1: 33–38.

—— (1999) "Management Behavior and Barriers to Market Orientation in Retailing Companies," *Journal of Services Marketing*, 13, 2: 113–131.

Harryson, S.J. (2002) *Managing Know-Who Based Companies.* (2nd edn, 1st edn in 2000), Cheltenham: Edward Elgar.

Hart, S., Tzokas, N. and Saren, M. (1999) "The Effectiveness of Market Information in Enhancing New Product Success Rates," *European Journal of Innovation Management*, 2, 1: 20–35.

Hatch, M.J. (1997) *Organization Theory. Modern, Symbolic, and Postmodern Perspectives*, New York: Oxford University Press.

Haverila, M. (1995) *The Role of Marketing when Launching New Products into the International Markets: An Empirical Study in Finnish High-Technology Companies*, Tampere: Tampere University of Technology.

Hayes, R.H. and Abernathy, W.J. (1980) "Managing Our Way to Economic Decline," *Harvard Business Review*, 58, 4: 67–77.

Hedlund, G. and Ridderstråle, J. (1995) "International Development Projects. Key to Competitiveness, Impossible, or Mismanaged?," *International Studies of Management & Organization*, 25, 1–2: 158–184.

Heenan, D.A. and Perlmutter, H.V. (1979) *Multinational Organization Development*, Reading: Addison-Wesley Publishing Company.

Hillebrand, B. (1996) "Internal and External Cooperation in Product Development," *Proceedings of the 3rd International Product Development Conference*, Fontainebleau, France.

Hillis, D. (2002) "Stumbling into Brilliance," *Harvard Business Review*, 80, 8: 152.

Hills, S.B. and Sarin, S. (2003) "From Market Driven to Market Driving: An Alternate Paradigm for Marketing in High Technology Industries," *Journal of Marketing Theory and Practice*, 11, 3: 13–23.

Hise, R.T., O'Neal, L., Parasuraman, A. and McNeal, J.U. (1990) "Marketing/R&D Interaction in New Product Development: Implications for New Product Success Rates," *Journal of Product Innovation Management*, 7, 2: 142–155.

Hoeffler, S. (2003) "Measuring Preferences for Really New Products," *Journal of Marketing Research*, 40, 4: 406–420.

Hofer, C.W. and Schendel, D. (1978) *Strategy Formulation: Analytical Concepts*, St Paul: West Publishing Company.

Hogan, J., Hogan, R. and Busch, C.M. (1984) "How to Measure Service Orientation," *Journal of Applied Psychology*, 69, 1: 167–173.

Holden, N.J. (2002) *Cross-Cultural Management. A Knowledge Management Perspective*, Essex: Pearson Education.

Hollensen, S. (2001) *Global Marketing. A Market-Responsive Approach* (2nd edn, 1st edn in 1998 by Prentice Hall), Harlow: Pearson Education.

Holt, K. (1976) "Need Assessment in Product Innovation," *International Studies of Management & Organization*, 6, 4: 26–44.

Holt, K., Geschka, H. and Peterlongo, G. (1984) *Need Assessment. A Key to User-Oriented Product Innovation*, Chichester: John Wiley & Sons.

Homburg, C. and Pflesser, C. (2000) "A Multiple-Layer Model of Market-Oriented Organizational Culture: Measurement Issues and Performance Outcomes," *Journal of Marketing Issues*, 37, 4: 449–462.

Hrebiniak, L.G. and Joyce, W.F. (1985) "Organizational Adaptation: Strategic Choice and Environmental Determinism," *Administrative Science Quarterly*, 30, 3: 336–349.

Hu, M.Y. and Griffith, D.A. (1997) "Conceptualizing the Global Marketplace: Marketing Strategy Implications," *Marketing Intelligence & Planning*, 15, 3: 117–123.

Huber, G.P. (1985) "Temporal Stability and Response-Order Biases in Participant Descriptions of Organizational Decisions," *Academy of Management Journal*, 28, 4: 943–950.

Huber, G.P. and Power, D.J. (1985) "Research Notes and Communications. Retrospective Reports of Strategic-Level Managers: Guidelines for Increasing their Accuracy," *Strategic Management Journal*, 2, 6: 171–180.

Hultink, E.J., Griffin, A., Hart, S. and Robben, H.S.J. (1997) "Industrial New Product Launch Strategies and Product Development Performance," *Journal of Product Innovation Management*, 14, 4: 243–257.

Hurley, R.F. (1998) "Customer Service Behavior in Retail Settings: A Study of the Effect of Service Provider Personality," *Journal of the Academy of Marketing Science*, 26, 2: 115–127.

Hurmelinna, P., Blomqvist, K., Saarenketo, S. and Nummela, N. (2002) "Trust and Contracting in the ICT Sector Born Global Companies," *Proceedings of the 28th EIBA Conference*, Athens, Greece.

Hutcheson, J.M. (1984) "International Marketing Techniques for Engineers," *International Marketing Review*, 1, 4: 51–59.

Hyysalo, S. (2004) *Uses of Innovation. Wristcare in the Practices of Engineers and Elderly*, Helsinki: University of Helsinki, Department of Education.

Ingledew, D.K., Markland, D. and Sheppard, K.E. (2004) "Personality and Self-Determination of Exercise Behaviour," *Personality and Individual Differences*, 36, 8: 1921–1932.

Innovation in Small and Medium Firms (1982) Background Reports. Paris: OECD.

Itami, H. and Roehl, T.W. (1987) *Mobilizing Invisible Assets*, Cambridge: Harvard University Press.

Jassawalla, A.R. and Sashittal, H.C. (1998) "An Examination of Collaboration in High-Technology New Product Development Processes," *Journal of Product Innovation Management*, 15, 3: 237–254.

Jauch, L.R. and Glueck, W.F. (1988) *Strategic Management and Business Policy*, New York: McGraw-Hill.

Jaworski, B.J. and Kohli, A.K. (1993) "Market Orientation: Antecedents and Consequences," *Journal of Marketing*, 57, 3: 53–70.

Jaworski, B., Kohli, A.K. and Shay, A. (2000) "Market-Driven Versus Driving Markets," *Journal of the Academy of Marketing Science*, 28, 1: 45–54.

Johannessen, J.-A., Olaisen, J. and Olsen, B. (1999) "Managing and Organizing Innovation in the Knowledge Economy," *European Journal of Innovation Management*, 2, 3: 116–128.

Johanson, J. and Vahlne, J.-E. (1977) "The Internationalization Process of the Firm – A Model of Knowledge Development and Increasing Foreign Market Commitments," *Journal of International Business*, 8, 1: 23–32.

John, G., Weiss, A.M. and Dutta, S. (1999) "Marketing in Technology-Intensive Markets: Toward a Conceptual Framework," *Journal of Marketing*, 63, Special Issue: 78–91.

Johnson, G. and Scholes, K. (1993) *Exploring Corporate Strategy*, Hertfordshire: Prentice Hall.

Jokinen, A. (2000) *Lobbying as a Part of Business Management*, Lappeenranta: Lappeenranta University of Technology.

Jolly, V.K. (1997) *Commercializing New Technologies. Getting from Mind to Market*, Boston: Harvard Business School Press.

Jones, R. (1996) "Do Athletes Make Good Reps?," *Sales & Marketing Management*, 148, 11: 92–96.

Kaplan, S.M. (1999) "Discontinuous Innovation and the Growth Paradox," *Strategy & Leadership*, 27, 2: 16–21.

Kappel, T.A. (2001) "Perspectives on Roadmaps: How Organizations Talk About the Future," *Journal of Product Innovation Management*, 18, 1: 39–50.

Keegan, W.J. (1969) "Multinational Product Planning: Strategic Alternatives," *Journal of Marketing*, 33, 1: 58–62.

Khandwalla, P.N. (1977) *The Design of Organizations*, New York: Harcourt Brace Jovanovich.

Khurana, A. and Rosenthal, S.R. (1998) "Towards Holistic 'Front Ends' in New Product Development," *Journal of Product Innovation Management*, 15, 1: 57–74.

Kipnis, D., Schmidt, S.M., Swaffin-Smith, C. and Wilkinson, I. (1984) "Patterns of Managerial Influence: Shotgun Managers, Tacticians, and Bystanders," *Organizational Dynamics*, 12, 3: 58–67.

Kleinschmidt, E.J. and Cooper, R.G. (1988) "The Performance Impact of an International Orientation on Product Innovation," *European Journal of Marketing*, 22, 10: 56–71.

—— (1991) "The Impact of Product Innovativeness on Performance," *Journal of Product Innovation Management*, 8, 4: 240–251.

Kohli, A.K. and Jaworski, B.J. (1990) "Market Orientation: The Construct, Research Propositions, and Managerial Implications," *Journal of Marketing*, 54, 2: 1–18.

Kotro, T. (2000) *Media and Mediators in the Product Development Process. Media Usage and the Transformations of Everyday Experience*, Turku: Media Studies, University of Turku. Medusa Research Project, Academy of Finland. www.uiah.fi/~tkotro/tanja_kotro_mediator.pdf (accessed 29.6.2004).

Kotro, T. and Pantzar, M. (2002) "Product Development and Changing Cultural Landscapes – Is Our Future in 'Snowboarding?'", Design Issues, 18, 2: 30–45.

Kotter, J.P. (1990) "What Leaders Really Do," *Harvard Business Review*, 68, 3: 103–111.

Kotter, J.P. and Heskett, J.L. (1992) *Corporate Culture and Performance*, New York: The Free Press.

Kremer Bennett, J. and O'Brien, M.J. (1994) "The Building Blocks of the Learning Organization," *Training*, 31, 6: 41–49.

Kuhn, T.S. (1970) *The Structure of Scientific Revolutions* (2nd edn, 1st edn in 1962), International Encyclopedia of Unified Science, Vol. 2, No. 2. Chicago: The University of Chicago Press.

Kumar, N., Scheer, L. and Kotler, P. (2000) "From Market Driven to Market Driving," *European Management Journal*, 18, 2: 129–142.

Kärkkäinen, H. (2002) *Customer Need Assessment: Challenges and Tools for Product Innovation in Business-to-Business Organizations*, Lappeenranta: Lappeenranta University of Technology.

Kärkkäinen, H., Piippo, P., Puumalainen, K. and Tuominen, M. (2001) "Assessment of Hidden and Future Customer Needs in Finnish Business-to-Business Companies," *R&D Management*, 31, 4: 391–407.

Lafferty, B.A. and Hult, G.T.M. (2001) "A Synthesis of Contemporary Market Orientation Perspectives," *European Journal of Marketing*, 35, 1–2: 92–109.

LaPlaca, P.J. and Punj, G. (1989) "The Marketing Challenge: Factors Affecting the Adoption of High-Technology Innovations," in R.W. Smilor (ed.) *Customer-Driven Marketing. Lessons from Entrepreneurial Technology Companies*, Lexington: Lexington Books, pp. 91–107.

Larson, L.L., Bussom, R.S. and Vicars, W. (1986) "Proactive *versus* Reactive Manager: Is the Dichotomy Realistic?," *Journal of Management Studies*, 23, 4: 385–400.

Lawler, E.E. III and Rhode, J.G. (1976) *Information and Control in Organizations*, Pacific Palisades: Goodyear Publishing Company.

Lawrence, P.R. and Lorsch, J.W. (1969) *Organization and Environment* (1st edn in 1967 by Harvard University Press), Homewood: Richard D. Irwin.

Lawton, L. and Parasuraman, A. (1980) "The Impact of the Marketing Concept on New Product Planning," *Journal of Marketing*, 44, 1: 19–25.

Leifer, R., O'Connor, G.C. and Rice, M. (2001) "Implementing Radical Innovation in Mature Firms: The Role of Hubs," *Academy of Management Executive*, 15, 3: 102–113.

Lemak, D.J. and Arunthanes, W. (1997) "Global Business Strategy: A Contingency Approach," *Multinational Business Review*, 5, 1: 26–38.

Leonard-Barton, D. (1990) "A Dual-Methodology for Case Studies. Synergistic Use of a Longitudinal Single Site With Replicated Multiple Sites," *Organization Science*, 1, 3: 248–266.

—— (1995) *Wellsprings of Knowledge. Building and Sustaining the Sources of Innovation*, Boston: Harvard School Press.

Leonard, D. and Rayport, J.F. (1997) "Spark Innovation through Empathic Design," *Harvard Business Review*, 75, 6: 102–113.

Lettl, C., Herstatt, C. and Gemünden, H.G. (2006) "Learning from Users for Radical Innovation," *International Journal of Technology Management*, 33, 1: 25–45.

Levitt, T. (1960) "Marketing Myopia," *Harvard Business Review*, 38, 4: 45–56.

—— (1963) "Creativity is Not Enough," *Harvard Business Review*, 8, 3: 72–83.

—— (1983) "The Globalization of Markets," *Harvard Business Review*, 63, 3: 92–102.

Li, T. and Cavusgil, S.T. (1999) "Measuring the Dimensions of Market Knowledge Competence in New Product Development," *European Journal of Innovation Management*, 2, 3: 129–145.

Li, T., Nicholls, J.A.F. and Roslow, S. (1999) "The Relationships Between Market-Driven Learning and New Product Success in Export Markets," *International Marketing Review*, 16, 6: 476–503.

Lieberman, M.B. and Montgomery, D.B. (1988) "First-Mover Advantages," *Strategic Management Journal*, 9, Special Issue: 41–58.

—— (1998) "First-Mover (Dis)advantages: Retrospective and Link with the Resource-Based View," *Strategic Management Journal*, 19, 12: 1111–1125.

Liedtka, J. (2000) "Strategic Planning as a Contributor to Strategic Change: A Generative Model," *European Management Journal*, 18, 2: 195–206.

Lincoln, Y.S. and Guba, E.G. (1985) *Naturalistic Inquiry*, Beverly Hills: Sage Publications.

Lindqvist, M., Sölvell, Ö. and Zander, I. (2000) "Technological Advantage in the International Firm – Local and Global Perspectives on the Innovation Process," *Management International Review*, 40, Special Issue: 95–126.

Liuhto, K. (1999) "The Impact of Environmental Stability on Strategic Planning: An Estonian Study," *International Journal of Management*, 16, 1: 98–111.

Løwendahl, B. and Revang, Ø. (1998) "Challenges to Existing Strategy Theory in a Postindustrial Society," *Strategic Management Journal*, 19, 8: 755–773.

Lukka, K. (1991) "Laskentatoimen tutkimuksen epistemologiset perusteet," *Liiketaloudellinen aikakauskirja*, 40, 2: 161–186.

Lukka, K. and Kasanen, E. (1995) "The Problem of Generalizability: Anecdotes and

Evidence in Accounting Research," *Accounting, Auditing & Accountability Journal*, 8, 5: 71–90.

Luo, Y. (1999) "Environment–Strategy–Performance Relations in Small Businesses in China: A Case of Township and Village Enterprises in Southern China," *Journal of Small Business Management*, 37, 1: 37–52.

Lynn, G.S., Morone, J.G. and Paulson, A.S. (1996) "Marketing and Discontinuous Innovation: The Probe and Learn Process," *California Management Review*, 38, 3: 8–37.

McCabe, D.L. and Dutton, J.E. (1993) "Making Sense of the Environment: The Role of Perceived Effectiveness," *Human Relations*, 46, 5: 623–643.

McDermott, C. and O'Connor, G.C. (2002) "Managing Radical Innovation: An Overview of Emergent Strategy Issues," *Journal of Product Innovation Management*, 19, 6: 424–438.

McGrath, J.E. (1982) "Dilemmatics. The Study of Research Choices and Dilemmas," in J.E. McGrath, J. Martin and R.A. Kulka (eds) *Judgement Calls in Research*, Beverly Hills: Sage Publications, pp. 69–102.

McGrath, R.G. and MacMillan, I.C. (1995) "Discovery-Driven Planning," *Harvard Business Review*, 73, 4: 44–54.

McKee, D.O., Varadarajan, P.R. and Pride, W.M. (1989) "Strategic Adaptability and Firm Performance: A Market-Contingent Perspective," *Journal of Marketing*, 53, 3: 21–35.

McKelvey, B. (1999) "Avoiding Complexity Catastrophe in Coevolutionary Pockets: Strategies for Rugged Landscapes," *Organization Science*, 10, 3: 294–321.

McKenna, R. (1989) "Why High-Tech Products Fail," in R.W. Smilor (ed.) *Customer-Driven Marketing: Lessons from Entrepreneurial Technology Companies*, Lexington: Lexington Books, pp. 3–14.

McKinnon, J. (1988) "Reliability and Validity in Field Reserach: Some Strategies and Tactics," *Accounting, Auditing and Accountability Journal*, 1, 1: 34–54.

Majaro, S. (1992) *Managing Ideas for Profit. The Creative Gap*, Berkshire: McGraw-Hill.

March, J.G. (1981) "Decision Making Perspective," in A.H. Van de Ven and W.F. Joyce (eds) *Perspectives on Organization Design and Behavior*, New York: John Wiley & Sons, pp. 205–244.

Marion, R. (1999) *The Edge of Organization. Chaos and Complexity Theories of Formal Social Systems*, Thousand Oaks: Sage Publications.

Markides, C. (1999) "Six Principles of Breakthrough Strategy," *Business Strategy Review*, 10, 2: 1–10.

Mascarenhas, B. (1992) "Order of Entry and Performance in International Markets," *Strategic Management Journal*, 13, 7: 499–510.

Mason, R.O., McKenney, J.L. and Copeland, D.G. (1997) "An Historical Method for MIS Research: Steps and Assumptions," *MIS Quarterly*, 21, 3: 307–320.

Meldrum, M.J. and Millman, A.F. (1991) "Ten Risks in Marketing High-Technology Products," *Industrial Marketing Management*, 20, 1: 43–50.

Meyer, A.D. (1995) "Tech talk.," How Managers are Stimulating Global R&D Communication," in J. Drew (ed.) *Readings in International Enterprise*, London: Routledge, pp. 179–195.

Miles, I. (1994) "Innovation in Services," in M. Dodgson and R. Rothwell (eds) *The Handbook of Industrial Innovation*, Hants: Edward Elgar, pp. 243–256.

Miles, M.B. (1979) "Qualitative Data as an Attractive Nuisance: The Problem of Analysis," *Administrative Science Quarterly*, 24, 4: 590–601.

Miles, M.B. and Huberman, A.M. (1994) *Qualitative Data Analysis* (2nd edn), Thousand Oaks: Sage Publications.

Miles, R.E. and Snow, C.C. (1978) *Organizational Strategy, Structure, and Process*, New York: McGraw-Hill.

Miles, R.H. (1982) *Coffin Nails and Corporate Strategies*, Englewood Cliffs: Prentice-Hall.

Millier, P. (1999) *Marketing the Unknown. Developing Market Strategies for Technical Innovations*, Chichester: John Wiley & Sons.

Mintzberg, H. (1973) "Strategy-Making in "Three Modes," *California Management Review*, 16, 2: 44–53.

—— (1979) *The Structuring of Organizations*, London: Prentice-Hall International.

—— (1991) "Strategic Thinking as Seeing," in J. Näsi (ed.) *Arenas of Strategic Thinking*, Helsinki: Foundation for Economic Education, pp. 21–25.

—— (1993) "The Pitfalls of Strategic Planning," *California Management Review*, 36, 1: 32–47.

—— (1994) *The Rise and Fall of Strategic Planning*, Hertfordshire: Prentice Hall.

Mohr, J. (2001) *Marketing of High-Technology Products and Innovations*, Upper Saddle River: Prentice Hall.

Moore, G.A. (1999a) *Crossing the Chasm* (2nd edn, 1st edn in 1991 by HarperCollins Publishers), Oxford: Capstone Publishing Limited.

—— (1999b) *Inside the Tornado, Marketing Strategies from Silicon Valley's Cutting Edge* (1st published in 1995), New York: Harper Collins Publishers.

Moorman, C. and Rust, R.T. (1999) "The Role of Marketing," *Journal of Marketing*, 63, Special Issue: 180–197.

Morgan, G. (1989) *Creative Organization Theory. A Resourcebook*, Newbury Park: Sage Publications.

—— (1992) "Proactive management," in D. Mercer (ed.) *Managing the External Environment. a Strategic Perspective*, London: Sage Publications, pp. 24–37.

—— (1997) *Images of Organization*, Thousand Oaks: Sage Publications.

Mullins, J.W. and Sutherland, D.J. (1998) "New Product Development in Rapidly Changing Markets: An Exploratory Study," *Journal of Product Innovation Management*, 15, 3: 224–236.

Möller, K. and Rajala, A. (1999) "Organizing Marketing in Industrial High-Tech Firms: The Role of Internal Marketing Relationships," *Industrial Marketing Management*, 28, 5: 521–535.

Nadler, D.A. and Tushman, M.L. (1990) "Beyond the Charismatic Leader: Leadership and Organizational Change," *California Management Review*, 32, 2: 77–97.

Narver, J.C. and Slater, S.F. (1990) "The Effect of a Market Orientation on Business Profitability," *Journal of Marketing*, 54, 4: 20–35.

Narver, J.C., Slater, S.F. and MacLachlan, D.L. (2004) "Responsive and Proactive Market Orientation and New-Product Success," *Journal of Product Innovation Management*, 21, 5: 334–347.

Nayak, P.R. and Ketteringham, J.M. (1986) *Breakthroughs!* London: Mercury Books.

New Products Management for the 1980s (1982) Booz Allen & Hamilton: New York.

Nierenberg, G.I. (1982) *The Art of Creative Thinking*, New York: Simon & Schuster.

Nikali, H. (1998) *The Substitution of Letter Mail in Targeted Communication*, Helsinki: Helsinki School of Economics and Business Administration.

Nooteboom, B. (2000) *Learning and Innovation in Organizations and Economies*, Oxford: Oxford University Press.

Normann, R. (1973) *A Personal Quest for Methodology*, Stockholm: Scandinavian Institutes for Administrative Research.

Oakley, P. (1996) "High-Tech NPD Success Through Faster Overseas Launch," *European Journal of Marketing*, 30, 8: 75–91.

O'Connor, G.C. (1998) "Market Learning and Radical Innovation: A Cross Case Comparison of Eight Radical Innovation Projects," *Journal of Product Innovation Management*, 15, 2: 151–166.

O'Connor, G.C. and McDermott, C.M. (2004) "The Human Side of Radical Innovation," *Journal of Engineering and Technology Management*, 21, 1–2: 11–30.

O'Connor, G.C. and Rice, M.P. (2001) "Opportunity Recognition and Breakthrough Innovation in Large Established Firms," *California Management Review*, 43, 2: 95–116.

O'Connor, G.C. and Veryzer, R.W. (2001) "The Nature of Market Visioning for Technology-Based Radical Innovation," *The Journal of Product Innovation Management*, 18, 4: 231–246.

Odén, B. (1989) "Tiden och periodisering i historien," in P. Heiskanen (ed.) *Aika ja sen ankaruus*, Helsinki: Oy Gaudeamus Ab, pp. 130–142.

Odendahl, T. and Shaw, A.M. (2002) "Interviewing Elites," in J.F. Gubrium and J.A. Holstein (eds) *Handbook of Interview Research. Context & Method*, Thousand Oaks: Sage Publications, pp. 299–316.

Oesterle, M-J. (1997) "Time-Span until Internationalization: Foreign Market Entry as a Built-in-Mechanism of Innovations," *Management International Review*, Special Issue 37, 2: 125–149.

Ohmae, K. (1989) "Managing in a Borderless World," *Harvard Business Review*, 67, 3: 152–161.

Oktemgil, M. and Greenley, G. (1997) "Consequences of High and Low Adaptive Capability in UK Companies," *European Journal of Marketing*, 31, 7: 445–466.

Olkkonen, R. (2002) *On the Same Wavelength? A Study of the Dynamics of Sponsorship Relationships between Firms and Cultural Organizations*, Turku: Turku School of Economics and Business Administration.

Olson, E.L. and Bakke, G. (2001) "Implementing the Lead User Method in a High Technology Firm: A Longitudinal Study of Intentions Versus Actions," *Journal of Product Innovation Management*, 18, 6: 388–395.

Olson, E.M., Walker, O.C. Jr and Ruekert, R.W. (1995) "Organizing for Effective New Product Development: The Moderating Role of Product Innovativeness," *Journal of Marketing*, 59, 1: 48–62.

O'Reilly, C. (1989) "Corporations, Culture, and Commitment: Motivation and Social Control in Organizations," *California Management Review*, 31, 4: 9–25.

Ott, J.S. (1989) *The Organizational Culture Perspective*, Pacific Grove: Brooks/Cole Publishing Company.

Ottesen, G.G. and Grønhaug, K. (2002) "Managers' Understanding of Theoretical Concepts: The Case of Market Orientation," *European Journal of Marketing*, 36, 11–12: 1209–1224.

Ottum, B.D. and Moore, W.L. (1997) "The Role of Market Information in New Product Success/Failure," *Journal of Product Innovation Management*, 14, 3: 258–273.

Parameswaran, R. and Yaprak, A. (1987) "A Cross-National Comparison of Consumer Research Measures," *Journal of International Business Studies*, 18, 1: 35–49.

Parker, L.D. (1997) "Informing Historical Research in Accounting and Management: Traditions, Philosophies, and Opportunities," *Accounting Historians Journal*, 24, 2: 111–149.

Parsons, T. (1951) *The Social System*, London: Routledge & Kegan Paul.

Pascale, R.T. (1999) "Surfing the Edge of Chaos," *MIT Sloan Management Review*, 40, 3: 83–94.

Patton, M.Q. (1990) *Qualitative Evaluation and Research Methods* (2nd edn, 1st edn in 1980), Newbury Park: Sage Publications.

Pauwels, P. and Matthyssens, P. (2004) "The Architecture of Multiple Case Study Research in International Business," in R. Marschan-Piekkari and C. Welch (eds) *Handbook of Qualitative Research Methods for International Business*, Cheltenham: Edward Elgar, pp. 125–143.

Pearson, A.E. (2002) "Though-Minded Ways to Get Innovative," *Harvard Business Review*, 66, 3: 99–106.

Penrose, E.T. (1980) *The Theory of the Growth of the Firm* (2nd edn, 1st edn in 1959), Oxford: Basil Blackwell.

Perlmutter, F.D. and Cnaan, R.A. (1995) "Entrepreneurship in the Public Sector: The Horns of a Dilemma," *Public Administration Review*, 55, 1: 29–36.

Perry, C. (1998) "Processes of a Case Study Methodology for Postgraduate Research in Marketing," *European Journal of Marketing*, 32, 9–10: 785–802.

Pettigrew, A.M. (1988) "Introduction: Researching Strategic Change," in A.M. Pettigrew (ed.) *The Management of Strategic Change*, Oxford: Basil Blackwell, pp. 1–13.

—— (1990) "Longitudinal Field Research on Change: Theory and Practice," *Organization Science*, 1, 3: 267–292.

—— (1992) "The Character and Significance of Strategy Process Research," *Strategic Management Journal*, 13, 8: 5–16.

—— (1997) "What is a Processual Analysis?," *Scandinavian Journal of Management*, 13, 4: 337–348.

Pfeffer, J. and Salancik, G.R. (1978): *The External Control of Organizations. A Resource Dependence Perspective*, New York: Harper & Row Publishers.

Pihlanto, P. (1994) "The Action-Oriented Approach and Case Study Method in Management Studies," *Scandinavian Journal of Management*, 10, 4: 369–382.

Pinkerton, B. (2001) "Using Market Research to Drive Successful Innovative Products," in J. Mohr (ed.) *Marketing of High-Technology Products and Innovations*, Upper Saddle River: Prentice Hall, pp. 117–118.

Porter, M.E. (1980) *Competitive Strategy: Techniques for Analysing Industries and Competitors*, New York: The Free Press.

Porter, M.E. and Sölvell, Ö. (1999) "The Role of Geography in the Process of Innovation and the Sustainable Competitive Advantage of Firms," in A.D. Chandler Jr, P. Hagström and Ö. Sölvell (eds) *The Dynamic Firm. The Role of Technology, Strategy, Organization, and Regions*, Oxford: Oxford University Press, pp. 440–457.

Powell, T.C. (1992) "Organizational Alignment as Competitive Advantage," *Strategic Management Journal*, 13, 2: 119–134.

Powell, W.W., Koput, K.W., and Smith-Doerr, L. (1996) "Interorganizational Collaboration and the Locus of Innovation: Networks of Learning in Biotechnology," *Administrative Science Quarterly*, 41, 1: 116–145.

Proff, H. (2002) "Business Unit Strategies between Regionalization and Globalization," *International Business Review*, 11, 2: 231–250.

Quinn, J.B. (1979) "Technological Innovation, Entrepreneurship, and Strategy," *MIT Sloan Management Review*, 20, 3: 19–30.

Rackham, N. (1998) "From Experience: Why Bad Things Happen to Good New Products," *Journal of Product Innovation Management*, 15, 3: 201–207.

Ragin, C.C. (1987) *The Comparative Method. Moving Beyond Qualitative and Quantitative Strategies*, Berkeley: University of California Press.

Rajala, A. (1997) *Marketing High Technology. Organising Marketing Activities in Process Automation Companies*, Helsinki: Helsinki School of Economics and Business Administration.

Reiss, S., Wiltz, J. and Sherman, M. (2001) "Trait Motivational Correlates of Athleticism," *Personality and Individual Differences*, 30, 7: 1139–1145.

Rhodes, R.E., Courneya, K.S. and Jones, L.W. (2004) "Personality and Social Cognitive Influences on Exercise Behavior: Adding the Activity Trait to the Theory of Planned Behavior," *Psychology of Sport and Exercise*, 5, 3: 243–254.

Rice, M.P., O'Connor, G.C., Peters, L.S. and Morone, J.G. (1998) "Managing Discontinuous Innovation," *Research Technology Management*, 41, 3: 52–58.

Rickards, T. (1985) *Stimulating Innovation. A Systems Approach*, London: Frances Pinter.

Ridderstråle, J. (1996) *Global innovation. Managing international innovation projects at ABB and Electrolux*, Stockholm: Stockholm School of Economics.

Ritter, T. and Gemünden, H.G. (2003) "Network Competence: Its Impact on Innovation Success and its Antecedents," *Journal of Business Research*, 56, 9: 745–755.

Roberts, E.B. (1990) "Evolving Toward Product and Market-Orientation: The Early Years of Technology-Based Firms," *Journal of Product Innovation Management*, 7, 4: 274–287.

Robertson, I.T. and Smith, M. (2001) "Personnel Selection," *Journal of Occupational and Organizational Psychology*, 74, 4: 441–472.

Robertson, T.S. (1971) *Innovative Behavior and Communication*, New York: Holt, Rinehart and Winston.

Rogers, E.M. (1983) *Diffusion of Innovations* (3rd edn, 1st edn in 1962), New York: The Free Press.

Rollinson, D. (2002) *Organisational Behaviour and Analysis. An Integrated Approach* (2nd edn, 1st edn in 1998), Harlow: Pearson Education.

Roth, K. and Morrison, A.J. (1990) "An Empirical Analysis of the Integration-Responsiveness Framework in Global Industries," *Journal of International Business Studies*, 21, 4: 541–564.

Rowley, J. (1997) "Focusing on Customers," *Library Review*, 46, 2: 81–89.

—— (2002) "Using Case Studies in Research," *Management Research News*, 25, 1: 16–27.

Salavou, H. and Lioukas, S. (2003) "Radical Product Innovations in SMEs: The Dominance of Entrepreneurial Orientation," *Creativity and Innovation Management*, 12, 2: 94–108.

Sandberg, B. (2002) "Creating the Market for Disruptive Innovation: Market Proactiveness at the Launch Stage," *Journal of Targeting, Measurement & Analysis for Marketing*, 11, 2: 184–196.

—— (2004) "Creating an International Market for Disruptive Innovation: The Domosedan® Case," in N. Nummela (ed.) *In Search of Excellence in International Business. Essays in Honour of Professor Sten-Olof Hansén*, Turku: Turku School of Economics and Business Administration, pp. 71–95.

—— (2007a) "Customer-Related Proactiveness in the Radical Innovation Development Process," *European Journal of Innovation Management*, 10, 2: 252–267.

—— (2007b) "Enthusiasm in the Development of Radical Innovations," *Creativity and Innovation Management*, 16, 3.

Sandberg, B. and Hansén, S.-O. (2004) "Creating an International Market for Disruptive Innovations," *European Journal of Innovation Management*, 7, 1: 23–32.

Sauser, W.I. Jr. and Sauser, L.D. (2002) "Changing the Way We Manage Change," *S.A.M. Advanced Management Journal*, 67, 4: 34–39.

Savitt, R. (1980) "Historical Research in Marketing," *Journal of Marketing*, 44, 4: 52–58.

—— (1987) "Entrepreneurial Behavior and Marketing Strategy," in A.F. Firat, N. Dholakia and R.P. Bagozzi (eds) *Philosophical and Radical Thought in Marketing*, Lexington: Lexington Books, pp. 307–322.

Schein, E.H. (1984) "Coming to a New Awareness of Organizational Culture," *MIT Sloan Management Review*, 25, 2: 3–16.

Schuh, A. (2001) "Strategic Change During the Internationalisation of the Firm," *Proceedings of the 27th EIBA Conference*, Paris, France.

Schumpeter, J.A. (1927) "The Explanation of the Business Cycle," *Economica*, December, 21: 286–311.

—— (1934) *The Theory of Economic Development. An Inquiry into Profits, Capital, Credit, Interest, and the Business Cycle*, Cambridge: Harvard University Press.

Scott, W.G., Mitchell, T.R. and Birnbaum, P.H. (1981) *Organization Theory: A Structural and Behavioral Analysis* (4th edn, 1st edn in 1972), Homewood: Richard R. Irwin.

Shane, S. and Venkataraman, S. (2000) "The Promise of Entrepreneurship as a Field of Research," *Academy of Management Review*, 25, 1: 217–226.

Shanklin, W.L. and Ryans, J.K. Jr. (1984) "Organizing for High-tech Marketing," *Harvard Business Review*, 84, 6: 164–171.

Sheth, J.N. and Parvatiyar, A. (2001) "The Antecedents and Consequences of Integrated Global Marketing," *International Marketing Review*, 18, 1: 16–29.

Shoham, A., Rose, G.M. and Albaum, G.S. (1995) "Export Motives, Psychological Distance, and the EPRG Framework," *Journal of Global Marketing*, 8, 3–4, 9–37.

Shrivastava, P. (1985) "Integrating Strategy Formulation with Organizational Culture," *Journal of Business Strategy*, 5, 3: 103–111.

Silverman, D. (2001) *Interpreting Qualitative Data. Methods for Analysing Talk, Text and Interaction* (2nd edn, 1st edn in 1993), London: Sage Publications.

Simon, H.A. (1984) *The Sciences of the Artificial* (2nd edn, 1st edn in 1969), Cambridge: The MIT Press.

Simon, M., Elango, B., Houghton, S.M. and Savelli, S. (2002) "The Successful Product Pioneer: Maintaining Commitment while Adapting to Change," *Journal of Small Business Management*, 40, 3: 187–203.

Sinkula, J.M., Baker, W.E. and Noordewier, T. (1997) "A Framework for Market-Based Organizational Learning: Linking Values, Knowledge, and Behavior," *Journal of the Academy of Marketing Science*, 25, 4: 305–318.

Sirilli, G. and Evangelista, R. (1998) "Technological Innovation in Services and Manufacturing: Results from Italian Surveys," *Research Policy*, 27, 9: 881–899.

Slater, S.F. (2001a) "Market Orientation at the Beginning of a New Millennium," *Managing Service Quality*, 11, 4: 230–232.

—— (2001b) "Putting It All Together: Market Orientation, Innovation, Research, and Customers," in J. Mohr (ed.) *Marketing of High-Technology Products and Innovations*, Upper Saddle River: Prentice Hall, pp. 169–171.

Slater, S.F. and Mohr, J.J. (2006) "Successful Development and Commercialization of Technological Innovation: Insights Based on Strategy Type," *Journal of Product Innovation Management*, 22, 1: 26–33.

Slater, S.F. and Narver, J.C. (1995) "Market Orientation and the Learning Organization," *Journal of Marketing*, 59, 3: 63–74.

—— (1998) "Customer-Led and Market-Oriented: Let's Not Confuse the Two," *Strategic Management Journal*, 19, 10: 1001–1006.

—— (2000) "Intelligence Generation and Superior Customer Value," *Journal of the Academy of Marketing Science*, 28, 1: 120–127.

Slevin, D.P. and Covin, J.G. (1990) "Juggling Entrepreneurial Style and Organizational Structure – How to Get Your Act Together," *MIT Sloan Management Review*, 31, 2: 43–53.

Smart, C. and Vertinsky, I. (1984) "Strategy and the Environment: A Study of Corporate Responses to Crises," *Strategic Management Journal*, 5, 3: 199–213.

Smircich, L. and Stubbart, C. (1985) "Strategic Management in an Enacted World," *Academy of Management Review*, 10, 4: 724–736.

Solberg, C.A. (2002) "The Perennial Issue of Adaptation or Standardization of International Marketing Communication: Organizational Contingencies and Performance," *Journal of International Marketing*, 10, 3, 1–21.

Sommers, W.P. (1982) "Product Development: New Approaches in the 1980s," in M.L. Tushman and W.L. Moore (eds) *Readings in the Management of Innovation*, Marshfield: Pitman Publishing, pp. 51–59.

Song, X.M. and Montoya-Weiss, M.M. (1998) "Critical Development Activities for Really New versus Incremental Products," *Journal of Product Innovation Management*, 15, 2: 124–135.

Song, X.M., Montoya-Weiss, M.M. and Schmidt, J.B. (1997) "Antecedents and Consequences of Cross-Functional Cooperation: A Comparison of R&D, Manufacturing, and Marketing Perspectives," *Journal of Product Innovation Management*, 14, 1: 5–47.

Souder, W.E., Sherman, J.D. and Davies-Cooper, R. (1998) "Environmental Uncertainty, Organizational Integration, and New Product Development Effectiveness: A Test of Contingency Theory," *Journal of Product Innovation Management*, 15, 6: 520–533.

Spencer, J.W. (2003) "Firms' Knowledge-Sharing Strategies in the Global Innovation System: Empirical Evidence from the Flat Panel Display Industry," *Strategic Management Journal*, 24, 3: 217–233.

Spender, J.-C. (1989) *Industry Recipes: An Enquiry into the Nature and Sources of Managerial Judgement*, Oxford: Blackwell.

—— (1993) "Some Frontier Activities around Strategy Theorizing," *Journal of Management Studies*, 30, 1: 11–30.

Srinivasan, R., Lilien, G.L. and Rangaswamy, A. (2002) "Technological Opportunism and Radical Technology Adoption: An Application to E-Business," *Journal of Marketing*, 66, 3: 47–60.

Stacey, R.D. (1993) *Strategic Management and Organisational Dynamics*, London: Pitman Publishing.

Stake, R.E. (1995) *The Art of Case Study Research*, Thousand Oaks: Sage Publications.

Steele, M. (1992) "The Changing Business Environment," in D. Faulkner and G. Johnson (eds) *The Challenge of Strategic Management*, London: Kogan Page, pp. 23–36.

Stefik, M. and Stefik, B. (2006) *Breakthrough. Stories and Strategies of Radical Innovation*, Cambridge: The MIT Press.

Stubbart, C.I. (1989) "Managerial Cognition: A Missing Link in Strategic Management Research," *Journal of Management Studies*, 26, 4: 325–347.

Sull, D.N. (1999) "Why Good Companies Go Bad," *Harvard Business Review*, 77, 4: 42–52.

Sultan, F. and Barczak, G. (1999) "Turning Marketing Research High-Tech," *Marketing Management*, 8, 4: 24–30.

Szymanski, D.M., Bharadwaj, S.G. and Varadarajan, P.R. (1993) "Standardization Versus Adaptation of International Marketing Strategy: An Empirical Investigation," *Journal of Marketing*, 57, 4: 1–17.

Sölvell, Ö. and Zander, I. (1995) "Organization of the Dynamic Multinational Enterprise," *International Studies of Management and Organization*, 25, 1–2: 17–38.

Tan, J.J. and Litschert, R.J. (1994) "Environment – Strategy Relationship and Its Performance Implications: An Empirical Study of the Chinese Electronics Industry," *Strategic Management Journal*, 15, 1: 1–20.

Teece, D.J. (1992) "Competition, Cooperation, and Innovation. Organizational arrangements for regimes of rapid technological progress," *Journal of Economic Behavior and Organization*, 18, 1: 1–25.

—— (2000) *Managing Intellectual Capital*, Oxford: Oxford University Press.

Teece, D.J., Pisano, G. and Shuen, A. (1997) "Dynamic Capabilities and Strategic Management," *Strategic Management Journal*, 18, 7: 509–533.

Tellis, G.J. and Golder, P.N. (1996) "First to Market, First to Fail? Real Causes of Enduring Market Leadership," *MIT Sloan Management Review*, 37, 2: 65–75.

Thompson, J.D. (1974) *Miten organisaatiot toimivat*, [*Organizations in Action*] (Published in 1967 by McGraw-Hill), Helsinki: Weilin+Göös.

Thompson, P. (1988) *The Voice of the Past* (2nd edn, 1st edn in 1978), Oxford: Oxford University Press.

Tidd, J., Bessant, J. and Pavitt, K. (1997) *Managing Innovation. Integrating Technological, Market and Organizational Change*, Chichester: John Wiley & Sons.

Trice, H.M. and Beyer, J.M. (1984) "Studying Organizational Cultures Through Rites and Ceremonials," *Academy of Management Review*, 9, 4: 653–669.

Trommsdorff, V. (1995) "Vorwort und Einleitung," in V. Trommsdorff (ed.) *Fallstudien zum Innovationsmarketing*, München: Verlag Franz Vahlen, pp. 1–11.

Trott, P. (2002) *Innovation Management and New Product Development* (2nd edn, 1st edn in 1998), Harlow: Pearson Education.

Turtiainen, J. (1999) "Ennakointi – käsitteellistä analyysiä," *Futura*, 18, 2: 25–31.

Tushman, M.L. and Anderson, P. (1986) "Technological Discontinuities and Organizational Environments," *Administrative Science Quarterly*, 31, 3: 439–465.

Urban, G.L. and Hauser, J.R. (1993) *Design and Marketing of New Products* (2nd edn), Englewood Cliffs: Prentice-Hall.

Urban, G.L. and von Hippel, E. (1988) "Lead User Analyses for the Development of New Industrial Products," *Management Science*, 34, 5: 569–582.

Urban, G.L., Weinberg, B.D. and Hauser, J.R. (1996) "Premarket Forecasting of Really-New Products," *Journal of Marketing*, 60, 1: 47–60.

Uslay, C., Malhotra, N.K. and Citrin, A.V. (2004) "Unique Marketing Challenges at the Frontiers of Technology: An Integrated Perspective," *International Journal of Technology Management*, 28, 1: 8–30.

Utterback, J.M. and Suárez, F.F. (1993) "Innovation, Competition, and Industry Structure," *Research Policy*, 22, 1: 1–21.

Valdelin, J. (1974) *Productutveckling och marknadsföring. En undersökning av productutvecklingsprocesser i svenska företag*, Stockholm: Ekonomiska Forskningsinstitutet vid Handelshögskolan.

Vancouver, J.B. (1996) "Living Systems Theory as a Paradigm for Organizational Behavior: Understanding Humans, Organizations, and Social Processes," *Behavioral Science*, 41, 3: 165–205.

Van de Ven, A.H. (1988) "Review Essay: Four Requirements for Processual Analysis," in A.M. Pettigrew (ed.) *The Management of Strategic Change*, Oxford: Basil Blackwell, pp. 330–341.

Van de Ven, A.H. and Poole, M.S. (1990) "Methods for Studying Innovation Development in the Minnesota Innovation Research Program," *Organization Science*, 1, 3: 313–335.

Van de Ven, A.H., Polley, D.E., Garud, R. and Venkataraman, S. (1999) *The Innovation Journey*, New York: Oxford University Press.

Vealey, R.S. (1992) "Personality and Sport: A Comprehensive View," in T.S. Horn (ed.) *Advances in Sport Psychology*, Champaign: Human Kinetics Publishers, pp. 25–59.

Veldhuizen, E., Hultink, E.J. and Griffin, A. (2001) "A Market Information Model for the Development of New Intelligent Products," *Proceedings of the 8th International Product Development Management Conference*, Enschede, The Netherlands.

Venkatraman, N. and Prescott, J.E. (1990) "Environment–Strategy Coalignment: An Empirical Test of Its Performance Implications," *Strategic Management Journal*, 11, 1: 1–23.

Verganti, R. (1999) "Planned Flexibility: Linking Anticipation and Reaction in Product Development Projects," *Journal of Product Innovation Management*, 16, 4: 363–376.

Veryzer, R.W. Jr. (1998a) "Discontinuous Innovation and the New Product Development Process," *Journal of Product Innovation Management*, 15, 4: 304–321.

—— (1998b) "Key Factors Affecting Customer Evaluation of Discontinuous New Products," *Journal of Product Innovation Management*, 15, 2: 136–150.

Vesalainen, J. (1995) *The Small Firm as an Adaptive Organization. Organizational Adaptation Versus Environmental Selection within Environmental Change*, Vaasa: University of Vaasa.

Volberda, H.W. (1996) "Toward the Flexible Form: How to Remain Vital in Hypercompetitive Environments," *Organization Science*, 7, 4: 359–374.

—— (1998) *Building the Flexible Firm. How to Remain Competitive*, Oxford: Oxford University Press.

Volberda, H.W. and Rutges, A. (1999) "FARSYS: A Knowledge-Based System for Managing Strategic Change," *Decision Support Systems*, 26, 2: 99–123.

von Bertalanffy, L. (1973) *General systems theory* (1st published in the United States in 1968), Harmondsworth: Penguin University Books.

von Hippel, E. (1986) "Lead Users: A Source of Novel Product Concepts," *Management Science*, 32, 7: 791–805.

—— (1989) "New Product Ideas from 'Lead Users,'" *Research & Technology Management*, 32, 3: 24–27.

—— (2006) *Democratizing Innovation*, Cambridge: The MIT Press.

von Hippel, E., Thomke, S. and Sonnack, M. (2000) "Creating Breakthroughs at 3M," *Health Forum Journal*, 43, 4: 20–27.

Wack, P. (1992) "Scenarios: Shooting the Rapids," in D. Mercer (ed.) *Managing the External Environment. A Strategic Perspective*, London: Sage Publications, pp. 235–254.

Wah, L. (1999) "The Almighty Customer," *Management Review*, 88, 2: 16–22.

Ward, A., Liker, J.K., Cristiano, J.J. and Sobek, D.K. II (1995) "The Second Toyota Pradox: How Delaying Decisions Can Make Better Cars Faster," *MIT Sloan Management Review*, 36, 3: 43–61.

Warren, K., Franklin, C. and Streeter, C.L. (1998) "New Directions in Systems Theory: Chaos and Complexity," *Social Work*, 43, 4: 357–373.

Watkins, M.D. and Bazerman, M.H. (2003) "Predictable Surprises: The Disasters You Should Have Seen Coming," *Harvard Business Review*, 81, 3: 72–80.

Webster, C. (1993) "Refinement of the Marketing Culture Scale and the Relationship Between Marketing Culture and Profitability of a Service Firm," *Journal of Business Research*, 26, 2: 111–131.

Webster, F.E. Jr. (1994) "Executing the New Marketing Concept," *Marketing Management*, 3, 1: 9–16.

Weick, K.E. (1979) *The Social Psychology of Organizing* (2nd edn, 1st edn in 1969), New York: Random House.

—— (1988) "Enacted Sensemaking in Crisis Situations," *Journal of Management Studies*, 25, 4: 305–317.

Wernerfelt, B. (1984) "A Resource-Based View of the Firm," *Strategic Management Journal*, 5, 2: 171–180.

Wigand, R., Picot, A. and Reichwald, R. (1997) *Information, Organization and Management. Expanding Markets and Corporate Boundaries*, Chichester: John Wiley & Sons.

Willmott, H. (1990) "Beyond Paradigmatic Closure in Organizational Enquiry," in J. Hassard and D. Pym (eds) *The Theory and Philosophy of Oorganizations. Critical Issues and New Perspectives*, London: Routledge, pp. 44–60.

Wind, J. and Mahajan, V. (1997) "Issues and Opportunities in New Product Development: An Introduction to the Special Issue," *Journal of Marketing Research*, 34, 1: 1–12.

Wind, Y., Douglas, S.P. and Perlmutter, H.V. (1973) "Guidelines for Developing International Market Strategies," *Journal of Marketing*, 37, 2: 14–23.

Workman, J.P. Jr. (1993a) "Marketing's Limited Role in New Product Development in One Computer System's Firm," *Journal of Marketing Research*, 30, 4: 405–421.

—— (1993b) "When Marketing Should Follow Instead of Lead" *Marketing Management*, 2, 2: 8–19.

Wright, P., Kroll, M., Pray, B. and Lado, A. (1995) "Strategic Orientations, Competitive Advantage, and Business Performance," *Journal of Business Research*, 33, 2: 143–151.

Wyner, G.A. (1998/1999) "Rethinking Product Development," *Marketing Research*, 10, 4: 49–51.

—— (1999) "Anticipating Customer Priorities," *Marketing Research*, 11, 1: 36–38.

Yelkur, R. and Herbig, P. (1996) "Global Markets and the New Product Development Process," *Journal of Product & Brand Management*, 5, 6: 38–47.

Yin, R.K. (1981) "The Case Study Crisis: Some Answers," *Administrative Science Quarterly*, 26, 1: 58–65.

—— (1989) *Case Study Research. Design and Methods* (rev. edn, 1st published in 1984), Newbury Park: Sage Publications.

Yip, G.S. (1991) *Do American Businesses Use Global Strategy?* Working Paper, Report No. 91–101. Massachusetts: Marketing Science Institute.

Yu, J.H., Keown, C.F. and Jacobs, L.W. (1993) "Attitude Scale Methodology: Cross-Cultural Implications," *Journal of International Consumer Marketing*, 6, 2: 45–64.

Zander, I. and Sölvell, Ö. (2000) "Cross-Border Innovation in the Multinational Corporation," *International Studies of Management & Organization*, 30, 2: 44–67.

Zeithaml, C.P. and Zeithaml, V.A. (1984) "Environmental Management: Revising the Marketing Perspective," *Journal of Marketing*, 48, 2: 6–53.

Zinkhan, G.M. and Pereira, A. (1994) "An Overview of Marketing Strategy and Planning," *International Journal of Research in Marketing*, 11, 3: 185–218.

Zou, S. and Cavusgil, S.T. (1996) "Global Strategy: A Review and an Integrated Conceptual Framework," *European Journal of Marketing*, 30, 1: 52–69.

Index

Figures are indicated by bold page numbers, tables by italics.

Lightning Source UK Ltd.
Milton Keynes UK
UKOW051404050413

208707UK00002B/30/P